Carbene

Edited by Satyen Saha
and Arunava Manna

Published in London, United Kingdom

IntechOpen

Supporting open minds since 2005

Carbene
http://dx.doi.org/10.5772/intechopen.95696
Edited by Satyen Saha and Arunava Manna

Contributors
Satyen Saha, Arunava Manna, Sumit K. Panja, Abhineet Verma, Vittal B. Gudimetla, Bony P. Joy, Sudeep Paul, Nuriye Tuna Subaşı, Eman A. Ali, Varala Ravi, Kotra Vijay, Khandapu Bala Murali Krishna, Bollikolla Hari Babu, Kurra Mohan, Kempegowda Mantelingu, Yatheesh Narayana, Sandhya N. C., H.E. Dinesh, Sridhar B. Thimmaiah, Kanchugarakoppal S. Rangappa, Hideto Miyabe, Ruchi Bharti, Monika Verma, Ajay Thakur, Renu Sharma, Naceur Hamdi, Nevin Gürbüz, Ismail Özdemir, Khaireddine Dridi, Ichraf Slimani

First published in London, United Kingdom, 2022 by IntechOpen
IntechOpen is the global imprint of INTECHOPEN LIMITED, registered in England and Wales, registration number: 11086078, 5 Princes Gate Court, London, SW7 2QJ, United Kingdom
Printed in Croatia

British Library Cataloguing-in-Publication Data
A catalogue record for this book is available from the British Library

Additional hard and PDF copies can be obtained from orders@intechopen.com

Carbene
Edited by Satyen Saha and Arunava Manna
p. cm.
Print ISBN 978-1-83969-929-0
Online ISBN 978-1-83969-930-6
eBook (PDF) ISBN 978-1-83969-931-3

For EU product safety concerns:
IN TECH d.o.o., Prolaz Marije Krucifikse Kozulić 3, 51000 Rijeka, Croatia, info@intechopen.com or visit our website at intechopen.com.

Meet the editors

Dr. Satyen Saha received a BSc and MSc in Chemistry from Jadavpur University, India, in 1994 and 1996, respectively. He obtained a Ph.D. in Photochemistry from Hyderabad Central University, India, in 2002. He completed postdoctoral research at the University of Tokyo, Japan, and Georgetown University, Washington DC, USA. He received a postdoctoral fellowship and bridge fellowship for foreign researchers from the Japan Society for the Promotion of Science (JSPS). Currently, he is a professor in the Department of Chemistry, Institute of Science, Banaras Hindu University, India. His research interests include synthesis, structure (solid and liquid phases), Ionic liquids, NIR emitting materials, and photophysical studies (steady-state and time-resolved) of fluorescent molecules in condensed phases.

Dr. Arunava Manna is currently an assistant professor in the Department of Chemistry, Banaras Hindu University, India. He obtained his Ph.D. in Organic Chemistry from Carnegie Mellon University (CMU), Pennsylvania, USA. After graduating from CMU in 2015, he worked for a couple of years as a post-doctoral fellow at the University of Geneva, Switzerland, working on synthesizing peptide libraries for the screening of small molecules. Following this, Dr. Manna spent two years at Emory University, Georgia, USA, as a postdoc working on bio-catalysis. His research interests lie at the interface of organic chemistry and biology.

Contents

Preface XI

Chapter 1 1
Recent Development of Carbenes: Synthesis, Structure, Photophysical Properties and Applications
by Arunava Manna, Abhineet Verma, Sumit K. Panja and Satyen Saha

Chapter 2 29
Recent Advances in Cooperative N-Heterocyclic Carbene Catalysis
by Hideto Miyabe

Chapter 3 43
N-Heterocyclic Carbenes (NHCs): An Introduction
by Ruchi Bharti, Monika Verma, Ajay Thakur and Renu Sharma

Chapter 4 63
Basic Information about Carbenes
by Nuriye Tuna Subasi

Chapter 5 81
Late Transition Metal (LTM)-NHC Catalyzed Transformations of Renewable Chemicals to Fine Chemicals, Fuels, and Intermediates
by Kurra Mohan, Bollikolla Hari Babu, Khandapu Bala Murali Krishna, Kotra Vijay and Varala Ravi

Chapter 6 95
Novel N-Heterocyclic Carbene Silver (I) Complexes: Synthesis, Structural Characterization, Antimicrobial, Antioxidant and Cytotoxicity Potential Studies
by Ichraf Slimani, Khaireddine Dridi, Ismail Özdemir, Nevin Gürbüz and Naceur Hamdi

Chapter 7 113
Imidazolium-Based N-Heterocyclic Carbenes (NHCs) and Metal-Mediated Catalysis
by Vittal B. Gudimetla, Bony P. Joy and Sudeep Paul

Chapter 8 137
N-Heterocyclic Carbene Mediated Organocatalysis Reactions
by Yatheesh Narayana, Sandhya N.C., H.E. Dinesh, Sridhar B. Thimmaiah, Kanchugarakoppal S. Rangappa and Kempegowda Mantelingu

Chapter 9 157
N-Heterocyclic Carbenes: A Powerful Catalyst for Polymerization
by Eman A. Ali

Preface

Carbenes are important molecules in chemistry because of their photochemistry and high reactivity. They are usually short-lived, but some carbenes are highly stable, primarily stabilized by organometallic complexes. Generally, two types of carbenes are known: singlet carbene and triplet carbene. Both show unique photophysics, as well. Heterocyclic carbenes are currently emerging versatile ligands for organometallic complexes that can be used in medicinal and materials chemistry. For example, silver complexes with heterocyclic carbene complexes are found to exhibit antimicrobial and anticancer properties, while carbene-palladium complexes are known for their antitumor and antimicrobial activities. There are exciting reports on carbene complexes with rhodium, iridium, and platinum having intriguing biological activities. At present, Fischer carbene, Schrock carbene, and N-heterocyclic carbene (NHC) are the most popular carbenes. These carbenes are seen in applications such as cross-coupling, C-H, X-H insertion reactions, carbene catalyzed polymerization reactions, and more.

Over nine chapters, this book addresses important topics related to carbenes. The book begins with a chapter by Saha et al. that discusses the synthesis of important carbenes such as NHCs as well as their structural and photophysical studies.

Chapter 2 by Hideto reports on recent advances in cooperative NHC catalysis. The author highlights the application of several chiral NHC catalysts along with transition metal catalysts and photocatalysts to synthesize complex molecules. Chapter 3 by Bharti et al. focuses on the general structure and properties of carbene and NHC in particular. It discusses the unique properties of NHC and strategies for its synthesis. Chapter 4 by Tuna discusses the electronic and structural properties of carbene and carbene reactivity in organic synthesis. Chapter 5 by Varala et al. is on late transition metal (LTM)-NHC catalyzed transformations of renewable chemicals. The chapter also discusses the synthesis of biomass resources, fine chemicals, fuels, and intermediates with mechanistic aspects. Chapter 6 by Naceur et al. present an impressive review of novel NHC–silver (I) complexes. The authors discuss the synthesis and structural characterization of several derivatives along with studies of antimicrobial, antioxidant, and cytotoxicity potential. Chapter 7 by Babu is on imidazolium-based NHCs and metal-mediated catalysis. Chapter 8 by Mantelingu et al. examines NHC-mediated organocatalysis reactions and the non-umpolung activity of bis-electrophile α and β unsaturated acylazoliums reaction with suitable bis-nucleophiles in organic synthesis. Finally, Chapter 9 by Iman reports on the versatility of NHCs in polymerization as true organocatalysts for producing many industrial polymers. In addition, the chapter describes their use in step-growth polymerization to achieve high molecular weight polymers and as transesterification agents in ring-opening polymerization.

We thank all the contributors for their excellent chapters and Author Service Manager Ms. Maja Bozicevic at IntechOpen whose assistance made it possible for us to complete all reviews and publish this book on time.

Satyen Saha
Professor,
Department of Chemistry,
Institute of Science,
Banaras Hindu University,
India

Dr. Arunava Manna
Department of Chemistry,
Institute of Science,
Banaras Hindu University,
India

Chapter 1

Recent Development of Carbenes: Synthesis, Structure, Photophysical Properties and Applications

Arunava Manna, Abhineet Verma, Sumit K. Panja and Satyen Saha

Abstract

Carbenes are highly reactive intermediates in organic synthesis. These divalent carbon species are generally transient in nature and cannot be isolated. However, they can form stabile metal complexes. Later on, the development of N-heterocyclic carbene (NHC) and other stable carbene led to the application of these carbon (II) donor ligands in the synthesis of complex natural products, transition metal catalysis, organo-catalysis and several other synthetic methodologies. Here in this short review, we will discuss the brief history of the development of carbenes, synthesis of stable carbenes (NHC in particular), and their applications in natural products synthesis transition metal chemistry/organometallics. In addition to synthesis and application, the chapter will consist of a detailed structural analysis of carbenes and exciting photophysics of this class of compounds. Special emphasis will be given to electronic structure. The role of carbene in the development of luminescent NHC transition metal complexes, the tuning of emission properties as well as their active role as photocatalysts in the reduction of CO_2 will also be discussed.

Keywords: carbene, N-heterocyclic carbene, electronic structure, photophysics, applications

1. Introduction

Carbenes are divalent carbon compounds which are generally highly reactive organic intermediates with six valence electrons having the general formula R^2C or R^1R^2C [1, 2]. Carbenes are classified as either singlets or triplets, depending upon their electronic structure. Most carbenes are very short lived, although persistent carbene are also known.

Carbene generally have either linear (as an extreme case) or bent geometry with sp^2 hybridized central carbon atom. These structures are related to electronically different orbital coupled states of central carbon atom in carbene. The orbital coupling between sp-hybrid orbitals and other two energetically degenerated p-orbitals results linear geometry of carbene (**Figure 1**). In other case, the hybrid sp^2

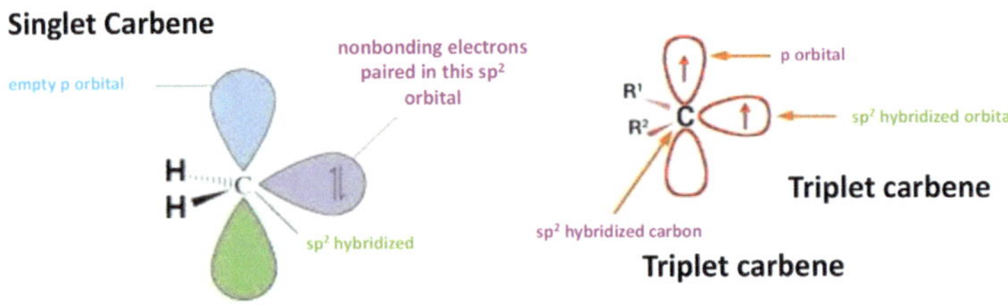

Figure 1.
Geometry (linear and bent) and hybridization of carbine; Single head arrow indicates electron.

orbitals (δ orbitals) coupled with a p orbital (p-π orbital) promotes bent geometry of carbine (**Figure 1**).

The arrangement of two nonbonding electrons in carbene is extremely important to reactivity of carbene. The two different electronic states are obtained from different arrangement of two nonbonding electrons in carbene. These electronic states are related to triplet and singlet state of carbene.

1.1 Triplet state of carbene

From two nonbonding electrons, one electron occupies in empty δ orbitals and another electron resides in empty pπ with parallel spin orientation ($\delta^1 p\pi^1$: 3B1). If electronic spin orientation is antiparallel, then the carbene is no longer triplet carbene ($\delta^1 p\pi^1$: 1B1) (**Figure 2**).

1.2 Singlet state of carbene

When the two nonbonding electrons occupy as a lone-pair in the emptyδ orbital, then the pπ orbital is being vacant ($\delta^2 p\pi^0$: 1A1 state). If these two nonbonding electrons are present in pπ orbital as a lone pair with empty δ-orbital, then 1A1 state is also created with $\delta^0 p\pi^2$electronic configuration. Interestingly, the $\delta^2 p\pi^0$ (1A1 state) is considered as a more stable state than the $\delta^0 p\pi^2$ (another 1A1 state) where the lone-pair occupies the pπ orbital (**Figure 2**).

The stability of singlet state ($\delta^2 p\pi^0$ (1A1 state)) is explained on the basis of significant energy difference between δ and pπ orbital (>2.0 eV).1 Singlet carbenes show the amphiphilic behavior (nucleophilic and electrophilic character) due to the presence of a sp2 hybridized lone-pair and of a vacant p-orbital.

Initially, carbene were so reactive that they were only considered as reaction intermediates or transition states. They could not be isolated and were only indirectly studied, often by trapping them in the presence of suitable reagents. However, now carbene can be stabilized and isolated by forming complexes with transition metals. They act as ligands for organometallic complexes. Two types of carbene-metal complex are known and they are Fischer and Schrock-type complexes (**Figure 3**).

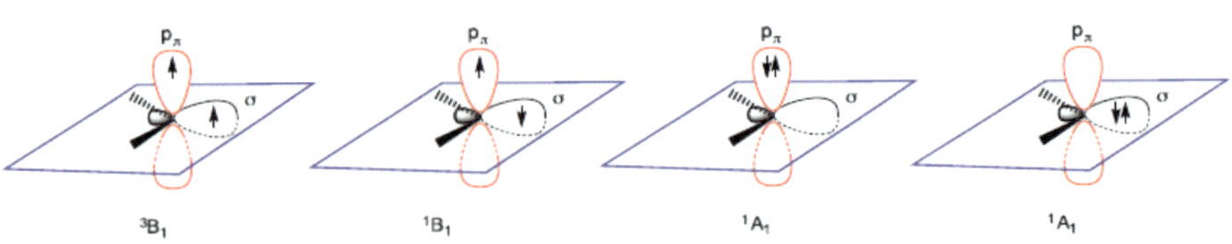

Figure 2.
Triplet and singlet carbenes: electronic configurations. Arrows: electrons.

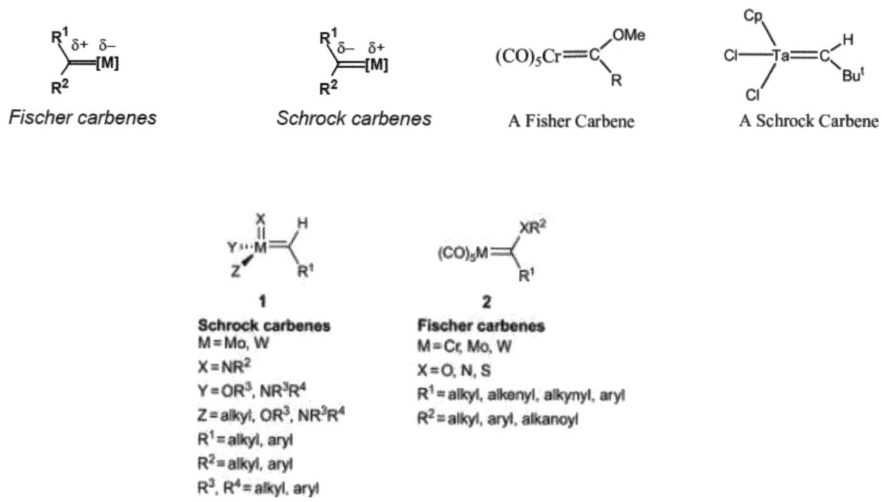

Figure 3.
General structure of Fisher and Schrock Carbene with examples.

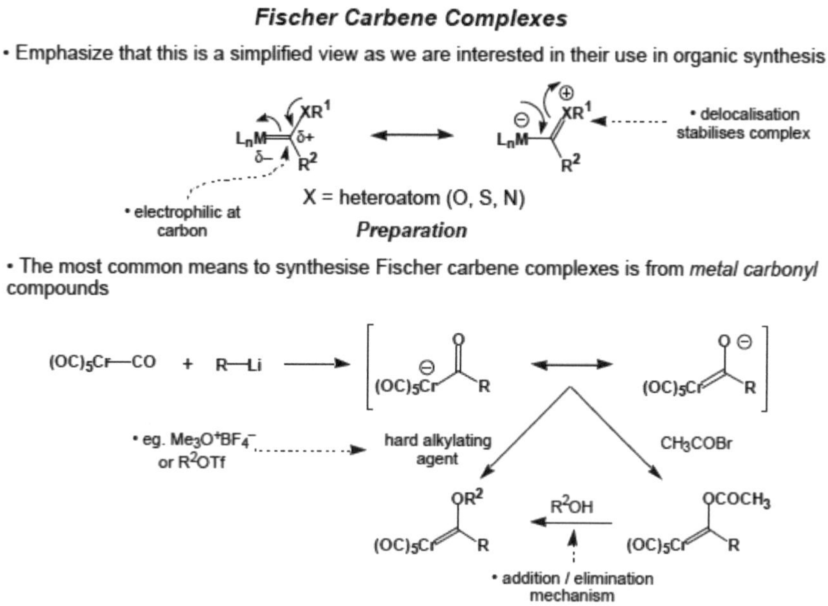

Figure 4.
Synthesis of Fischer carbene and their application in organic synthesis.

The Fischer carbene, which were first described in 1960s, form complexes with low valent or lower oxidation state of metal and are versatile reagents for organic synthesis due to presence of electrophilic carbon center (**Figure 4**). The Schrock-type compounds (first reported in the early 1970s) play an important role in olefin metathesis due to present of nucleophilic carbon center (**Figure 5**).

2. Types of N-heterocyclic carbene (NHC) ligands

The structure of carbene depends upon the nature of ligands.

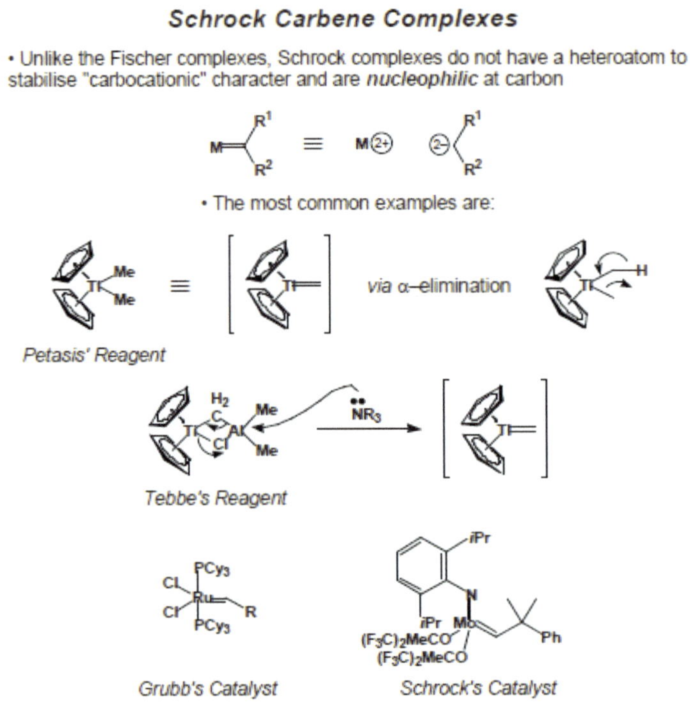

Figure 5.
Synthesis of Schrock's carbene and their application as catalyst.

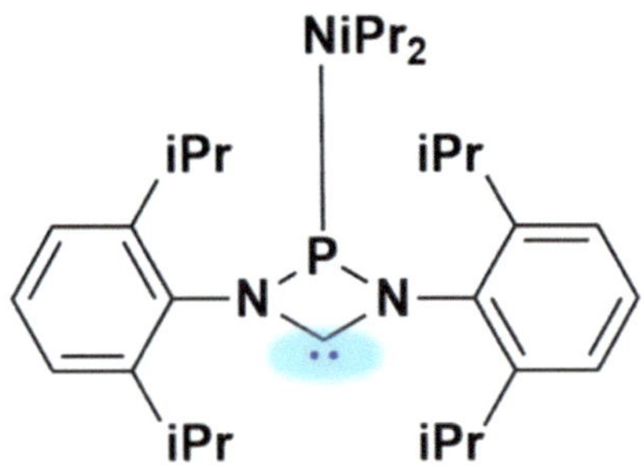

Figure 6.
A 4-membered N-heterocyclic carbene ligand 2,6-diisopropyl-substituted substituents.

2.1 4-Membered NHC

Grubbs et al. was the first group to develop the 4-membered NHC [3] (**Figure 6**). It was found that for the isolation of carbene carbon steric shielding was very much important. The 2, 6-diisopropyl-substituted constituents led to the successful isolation of the free NHC [3, 4]. The vibrational, ν(CO) values of the corresponding Rhodium dicarbonyl complex (ν(CO) in toluene: 2080 and 1988 cm^{-1}) show that it's σ-electron donating properties is slightly less than the dihydroimidazol-2-ylidene analogue [5].

2.2 5-Membered NHC

5-membered ring systems were most and widely reported NHC carbene so far [6–11]. This is due to the fact that 5-membered ring system is sterically more stable and hence provide the extra stability to the NHC as well as improving its catalytic

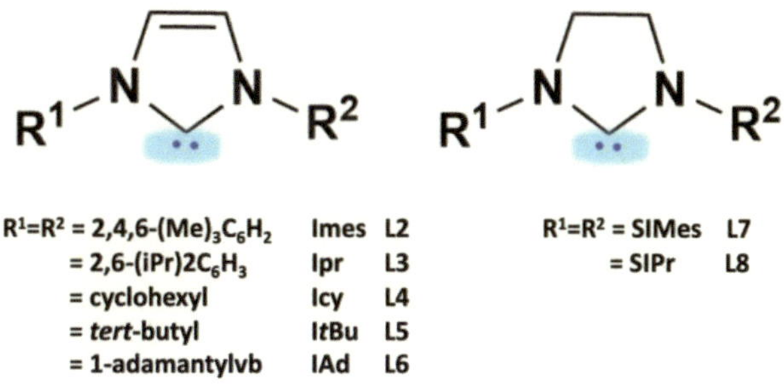

Figure 7.
Imidazol-2-ylidenes and imidazolidin-2-ylidenes.

properties [8, 9]. Some of the scaffolds used for the preparation of 5-membered NHC were imidazole-2-ylidenes Imes (**L2**), IPr (**L3**), Icy (**L4**), ItBu (**L5**), and IAd (**L6**) and the imidazolidin-2-ylidenes SIMes (**L7**) and SIPr (**L8**) presented in **Figure** 7.

IBlox system of NHC ligands have been well explored and readily derived from bioxazolines (**Figure 8**) (**L9–L13**) [5, 6] probably due to the two reasons. (1) The 4,5-dioxygen substitution affects the ligand's electronic properties. The electron donating capability is similar to electron-rich phosphines like $PtBu_3$, but slightly less electron-rich than other imidazolium-based N-heterocyclic carbene. It is fascinating that all IBiox ligands have the similar or same electronic properties. There are several salient features among the ligands (**L10–L13**):

a. all of them have a rigid tricyclic backbone

b. the presence of the substituents R1 and R2 on the peripheral rings. These groups shield the carbene carbon, and can cause the metal's coordination sphere either to expand or to contract.

c. the cycloalkyl substitution on the rigid tricyclic backbone enable the IBiox ligands to become sterically demanding, while being flexible at the same time (flexible steric bulk) [5–7].

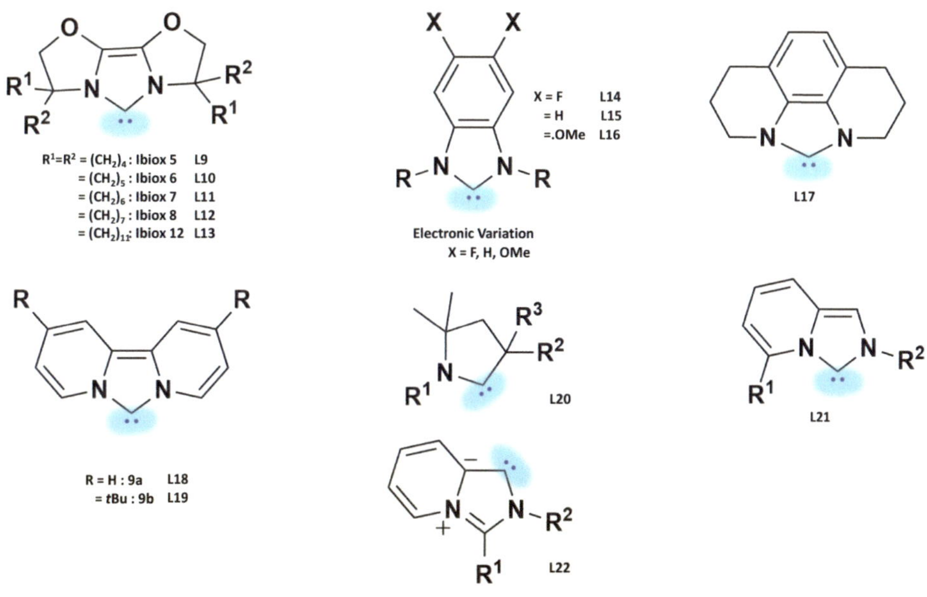

Figure 8.
Some of the most widely applicable 5-membered NHC.

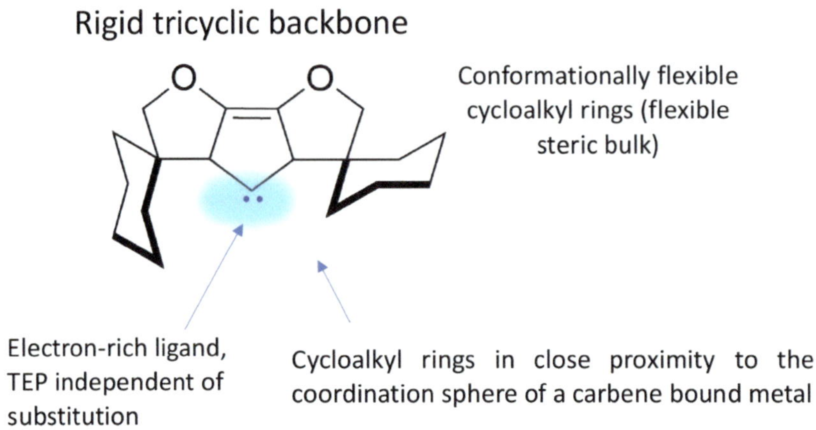

Figure 9.
Important feature for the IBiox NHC ligands.

A salient feature of these carbenes (NHC ligands) is that steric bulkiness of the ligands can be modified according to the uses without affecting the electronic character of carbene which is an ideal criterion for selection of ligands (**Figure 9**). It is an important and a unique property for such monodentate ligands as compared to monodentate phosphines where increasing the size of the phosphine ligands can affect both their steric and electronic properties.

N-heterocyclic carbene based on Benzimidazolium (**Figure 8**) **L14–L16** [12–16] and **L17** [17] are an important as well as interesting classes of carbenes, though less commonly explored classes of NHC. The synthetic challenges limited the scope to only three electronically different ligands (**L14–L16**) and no sterically tunable benzimidazolium-derived N-heterocyclic carbene [18, 19].

Weiss et al. was the first person to develop and introduce the Bipyridocarbene (**L18** and **L19**) which is a highly electron-rich NHC (**Figure 8**) [20, 21]. This is evident from the strong high-field shift of its carbene signal (196 ppm) in the ^{13}C NMR spectrum [22]. But the instability of this compound limited its application in catalysis. On the other hand, Kunz et al. showed that tert-butyl substitution can lead to the formation of more stable NHC (**L19**) and also reported for the first time, the X-ray structural analysis of these types of carbene [23]. Later on, Lassaletta et al. [8] and Glorius et al. [7] independently developed imidazo [1,5-a] pyridine-3-ylidenes. These can be viewed as benzannulated imidazolin-2-ylidenes **L2–L6.** These ligands form electron-rich carbene as seen in IR spectra. The ν(CO) for cis-(CO)2 RhCl with R1, R2 = Me was found to be 2079 and 2000 cm^{-1}.

The structures of some other interesting carbene ligands (**L20**) and (**L21**) based on imidazolium backbone are also shown. These ligands showed different reactivity in the palladium-catalysed α-arylation of propiophenone because of their structural features.

2.3 6- and 7-Membered NHC

6 or 7 membered ring carbenes of N-heterocyclic such as 1,3-disubstituted pyrimidin- 2-ylidenes **L23** [24–27], perimidine-based carbene **L24** [28], **L25–L27** [29] or chiral 7-membered NHC **L28** [30] have only rarely been reported (**Figure 10**). The different electronic properties of NHCs are due to the different backbone structures and in the topology of the substituents on the NHC. Richeson et al. validated this by incorporating a naphthyl ring system in ligand **L25**. This modification changed the shape of the NHC [28]. The value of the N-C carbene–N

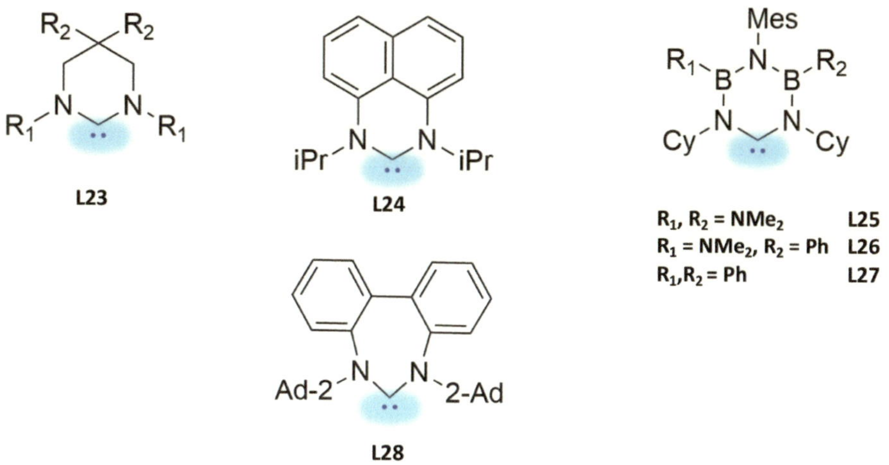

Figure 10.
Some most and widely applicable six and seven-membered NHC.

bond angle increased from 100 to 110° in 5 membered to 115.3° in 6 membered ring. The carbene N–R angle α is also reduced from 122 to 123° in (**L2–L6**) and (**L7** & **L8**) to 115.5° in **L24**, which had a steric influence of the N-substituents on the carbene carbon. Based on the ν(CO) values of the corresponding cis-(CO)2RhCl complex, ligand **L24** is an even stronger electron donor than the dihydroimidazol-2- ylidenes **L7** & **L8**, but weaker than the acyclic carbene $C(NiPr_2)_2$.

Borazines, also known as "inorganic benzene" and isoelectronic with benzene are excellent scaffolds for highly stable heterocycles. When the borane moiety is "exchanged" with an iso-electronic carbene moiety one can obtain NHC **L25–L27**. There have been reports on the synthesis of such stable complexes of these ligands but their catalytic properties have been not explored yet.

The first synthesis of a 7-membered NHC ligand was reported by Stahl et al. very recently [29, 30]. Even though NHC **L28** could not be isolated as a free carbene, palladium complexes of **L28** were isolated and their structures fully characterized. Ligand **L28** is C2 symmetric and because of a torsional twist it shows the Möbius-aromatic character of the 8π-electron carbene heterocycles [31].

2.4 Bi- and multi-dentate NHC

In addition to these monodentate ligands, several multi-dentate ligands have been synthesized and used for various applications. The rigid bidentate benzimidazole-based N-heterocyclic carbene was used in the synthesis of conjugated organometallic polymers which show interesting electronic and mechanical properties [32]. Another application of such bidentate NHC was the formation of stable chelate complexes. One such palladium-NHC complex was used in the catalytic conversion of methane to methanol [33]. The stability of the complexes is a pre-requirement for such applications as the reaction takes place in an acidic medium (trifluoroacetic acid) at high temperatures (80°C) in the presence of strong oxidizing agents like potassium peroxodisulfate.

Similar stable metal-chelate complexes were reported using tri- and tetra-dentate ligands. Iron (III) and chromium (III) form complexes of the structure $[M(L29)_2]^+$ with the tripodal tricarbene ligand **L29** (**Figure 11**) [34].

On the other hand, the development of macrocyclic ligands was found to be challenging. Hahn et al. successfully synthesized tetracarbene ligands having crown

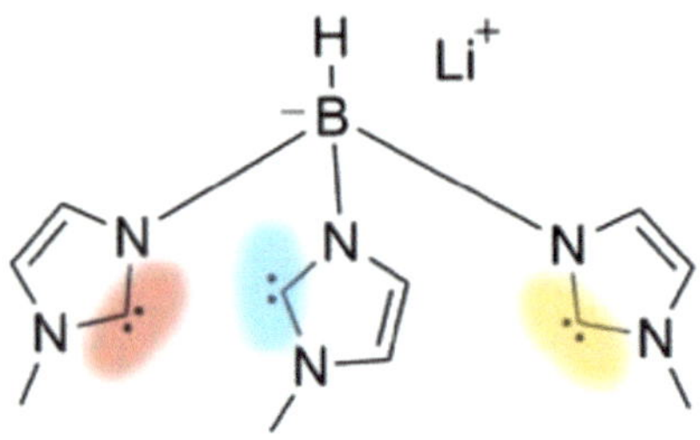

Figure 11.
*A tridentate ligand (**L29**).*

ether topology in a template-controlled synthetic approach.Initially, a transition metal complex with four unsubstituted benzimidazol-derived NHC **L14** to **L16** (R, X = H) was formed. The carbene ligands are not stable when separated from the transition metal. Then, the carbene ligands were linked by a template-controlled cyclization of alkyl or aryl isocyanides and finally, the desired product was synthesized.

3. Synthesis

While the first report on the synthesis of carbene dates back to the 1920s, N-Hetero carbene (NHC) was discovered, from independent research works by Ofele [35], Wanzlick [36], and Lappert [37] in 1960s. A (phosphino)-(silyl) carbene was the first stable carbene to be synthesized which led to the tremendous fluorishment in carbene chemistry (**Figure 12**: left) [36, 38, 39]. After long time from discovery of carbene, Arduengo et al. reported for the first time the isolation of metal-free N-heterocyclic carbene in 1991 (**Figure 12**: right) [40–44].

After the synthesis and isolation of stable free NHC, carbene chemistry has attracted much more attention in recent times and scientists began to look for new NHC ligands and the synthesis of stable NHC. Prior to this carbene were thought to be highly reactive to be isolated and thus it limited the studies on carbene. But the stability, isolation of NHC and the ease of synthesis from cheap and easily available precursors such as imidazolium salts, made the field advance rapidly in the last three decades. The stability of Arduengo-type free carbene owes to the presence of two heteroatoms in the molecule. The inductive effect of these heteroatoms stabilize the carbene [45]. The chemistry, structure, and propertiesof these "classical" heterocyclic carbene have been reviewed elaborately [46, 47]. The diverse applications of NHC have prompted the design and development of novel NHC structures. However, NHC are obtained generally from their suitable precursors. Thus, the facile, diverse synthesis of the NHC precursors is of great importance in order to get

Figure 12.
First isolated stable carbene compound.

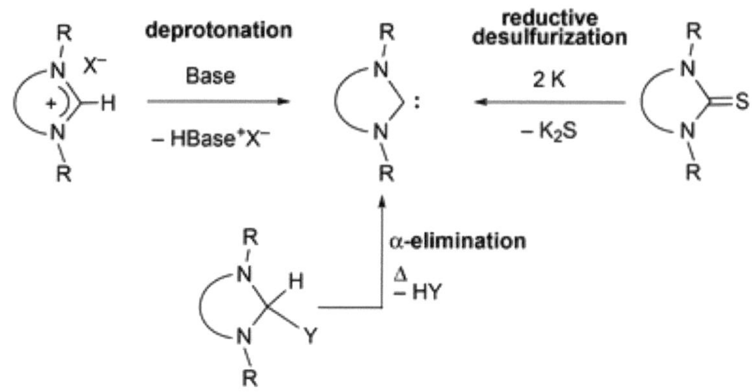

Figure 13.
Different strategies of carbene synthesis.

NHCs of various designs. Depending on NHC precursors their synthesis can be divided into the following major categories (**Figure 13**):

a. Deprotonation of azolium salts

b. Elimination reactions from imidazolines

c. Desulfurization of imidazol- and benzamidazol-2-thiones

Two of the most commonly used NHC scaffolds are imidazolin-2-ylidene and its saturated version, imidazolidin-2-ylidene. In both the cases, the two nitrogen atoms are substituted with alkyl and aryl groups either in a symmetrical or in an asymmetrical way [37]. One of the easiest and widely used methods for the preparation of NHCs is the deprotonation of imidazolium or imidazolinium salts with strong bases such as sodium hydride [38], potassium tert-butoxide [39], or potassium bis (trimethylsilyl)amide [40].

The azolium salts can be also be prepared by the following routes namely,

3.1 N-alkylation of heterocycles

The first approach is simple and straightforward; the successive alkylation of the nitrogen atom in these heterocycles generates the quaternary N atoms which are excellent NHC precursors. The various methodologies for synthesis of imidazole, oxazole, thiazole, and other five membered heterocyclic rings have been reviewed extensively [48–53].

3.2 Symmetrical synthesis of imidazolinium salts as NHC precursors

Symmetrical synthesis of imidazolinium salts as NHC precursors can be achieved by various methods. One such example is shown in **Figure 14** [54] following condensation reduction route. This is widely applicable for a variety of primary amines [55–64].

3.3 Unsymmetrical synthesis of imidazolinium salts as NHC precursors

Unsymmetrical synthesis of imidazolinium salts as NHC precursors similarly have been prepared following different methods. The uses of oxalyl chloride or

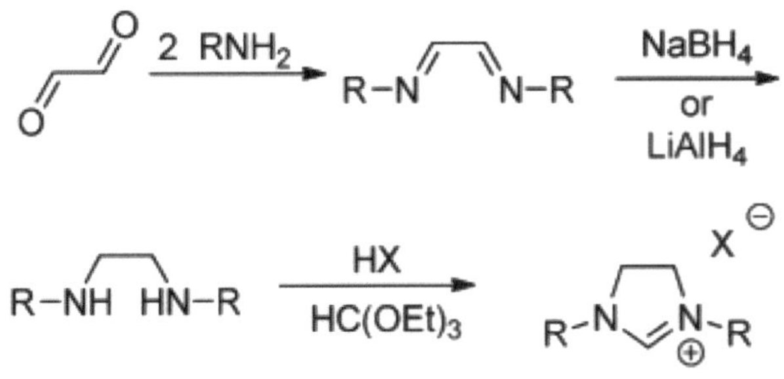

Figure 14.
Synthesis of symmetrical NHC precursor.

derivatives give more flexibility leading to the synthesis of unsymmetrical imidazoliniums. Mol and co-workers synthesized the mixed adamantyl/mesityl N-heterocyclic precursor 1-(1-adamantyl)-3- mesityl-imidazolinium chloride. At first, oxalyl chloride was reacted with mesityl amine to afford the intermediate acyl chloride which on successive reaction with another amine afforded the desired compound **(Figure 15)** [65, 66]. A small library of imidazolinium derivatives were preparedhaving groups with various steric or electronic properties using this method [67–72].

Similarly, the unsaturated azolinium salts have been prepared by (a) alkylation of the nitrogen atom of imidazolinium, (b) symmetric and (c) unsymmetrical synthesis of such salts.

Paraformaldehyde has been extensively used for the synthesis of both unsaturated and saturated NHC precursors as shown in **Figure 16** [73]. There are several other strategies for the preparation of such azolium salts as NHC precursors which have been nicely reviewed [54]. The deprotonation method of generating carbene

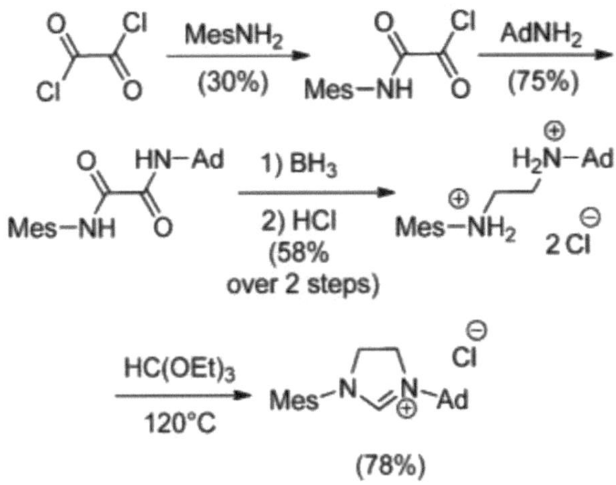

Figure 15.
Synthesis of unsymmetrical NHC precursor.

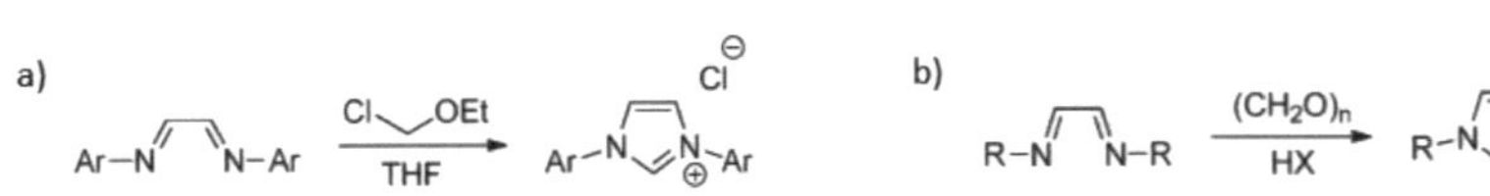

Figure 16.
Synthesis of unsaturated azolinium based NHC precursors.

from imidazolium salts using a strong base is generally performed in-situ so that the air sensitive free or ligated carbene is not isolated. But sometimes, the use of such a strong base and harsh reaction conditions leads to unwanted side-reactions. Thus to avoid those unnecessary complications, elimination of hydrogen atoms from imidazolinium salts are undertaken to provide NHC carbene.

Alternatively, less common routes involve reduction of thiourea derivatives, pyrolysis of an NHC – volatile compound adduct or release of NHC.

3.3.1 NHC·CO_2 zwitterions

The carbon dioxide adducts of NHCs can be prepared easily by passing carbon dioxide gas into a free carbene solution, and then evaporating the solvent. They are comparatively air stable and can be stored for a long time. Imidazol(in)ium-2-carboxylate derivatives act as an excellent NHC precursor for the synthesis of NHC-transition metal complexes by releasing carbon dioxide during thermolysis. The steric effect on N substituent and other electronic, steric factors affecting the stability of such precursors has been also studied [74]. Later on, imidazol(in)ium hydrogen carbonates have been shown as another excellent source of NHCs when they lose H_2CO_3 upon heating [75–79].

3.3.2 NHC – metal adducts

Another important NHC precursor is silver (I) complexes of NHC. There have been several types of such complexes like: Imidazolin-2-ylidene involving imidazole ring with substituents at the nitrogen atoms, Benzimidazol-2-ylidenes having a benzene ring fused with the imidazole moiety, imidazolidin-2-ylidenes- and related heterocycles. The first Ag (I)–NHC adduct was reported by Arduengo in the early 90s by the reaction of Ag(I) salt with a free NHC [80]. Later, the uses of Ag(OAc) and Ag_2O as silver base were reported to synthesize various Ag(I)–NHCs. These silver bases are used for the deprotonation of azolium salts, and generation of Ag(I)–NHCs in situ. These Ag(I)-NHC complexes easily decompose under thermolysis to provide the free carbenes for various applications. One limitation to this approach is the use of silver metal in stoichiometric amounts. The different synthetic routes as well the structural diversity and the applications of such precursors have been well established in the carbene literature [81].

3.3.3 Other NHC precursors

In addition to the above mentioned adducts there are reports of similar complexes of NHC namely the chloroform and pentafluorobenzene adducts of 1,3-dimesitylimidazolidin-2-ylidene (SIMes). They are stable at room temperature and afford the corresponding NHC on thermolysis [82, 83]. One such example is SIMes (H)(O-t-Bu) which can produce the corresponding NHC at room temperature. The alcohol adducts of triazolin-5-ylidene and imidazolidin-2-ylidene also proved to be excellent NHC precursors [84, 85] (**Figure 17**).

4. Photophysical studies of N-heterocyclic carbene

New efficient light-emitting materials related to iridium(III) and platinum(II) complexes have attracted research area and wide range of applications in OLED and WOLED technologies. The suitable ligand based iridium(III) and platinum(II)

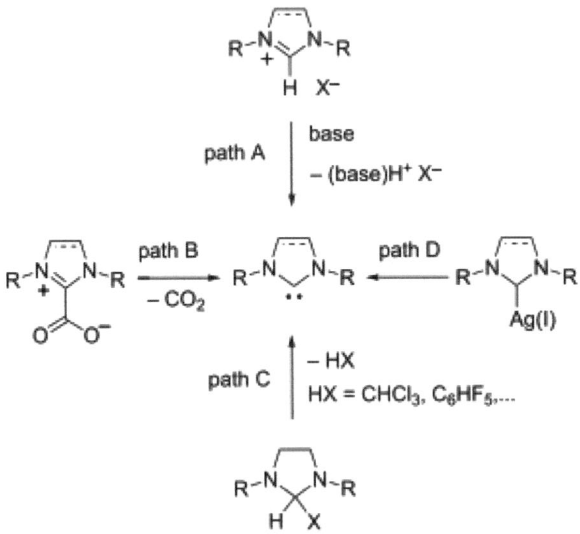

Figure 17.
Various strategies for generation of carbene.

complexes allow tailoring the emission properties for specific application in organic light emitting devices (OLEDs) and white organic light emitting devices (WOLEDs).

In particular, extensive investigations have been carried out on iridium(III) and platinum(II) complexes as triplet emitters in OLEDs. In OLEDs, significant progress has been achieved for making highly efficient and stable green and red emitters. But further advancement in recent progress of solid state full-colored OLED displays and WOLEDs appliances is also required in the research area of blue and white light emitting iridium(III) and platinum(II) complexes.

4.1 Photophysical studies of N-heterocyclic carbene platinum(II) complexes

Here N-heterocyclic carbene platinum(II) complexes are selected to investigated their photophysical properties (**Figure 18**). Selected N-heterocyclic carbene platinum(II) complexes have shown distinct absorption bands in 325–405 nm region with higher extinction coefficients (order of 10^3 or 10^4). These complexes are known as either blue, bluish green, or green emitters depending on emission bands within the 430–530 nm region with large Stokes shifts.

The $[Pt^{II}(C^\wedge N^\wedge C)Cl][PF_6]$ complex shows strong absorption bands at 272 nm and moderately intense bands at364 nm with higher extinction coefficients ($\varepsilon \sim 10^3$–10^4) in acetonitrile (**Figure 19**). The high-energy intense absorption band (277–291 nm) is assigned as $\pi \rightarrow \pi^*$ transitions (IL: Intra ligand) of the $C \equiv CR$ and C^N^C pincer ligands, whereas the low-energy absorption band (band (383–471 nm)) is observed due to presence of the $d\pi(Pt) \rightarrow \pi^*$ (C^N^C)] transitions, and $[\pi(C \equiv CR) \rightarrow \pi^*$ (C^N^C)] transition, considered as metal-to-ligand charge-transfer (MLCT), ligand-to-ligand charge-transfer (LLCT), mixed with the $\pi \rightarrow \pi^*$ transitions (IL) of the C^N^C pincer ligands [75–79].

A blue-shifted absorption band (383 nm) of Complex **1** with the alkylalkynyl ligand is observed compared to Complex 3 (appeared at 405 nm) with the phenylalkynyl because of weak π-donating ability of alkylalkynyl ligand. The

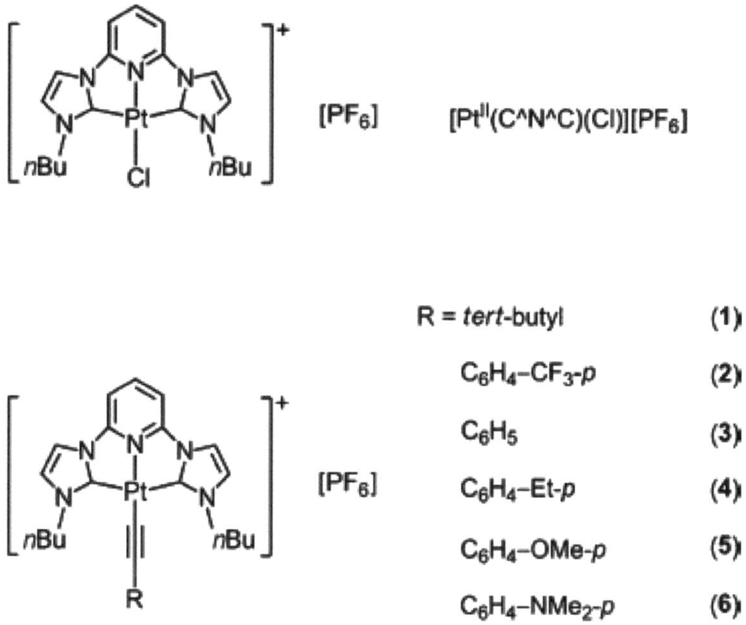

Figure 18.
Structures of pyridine-based N-Heterocyclic Carbene Platinum(II) Complexes (1–6).

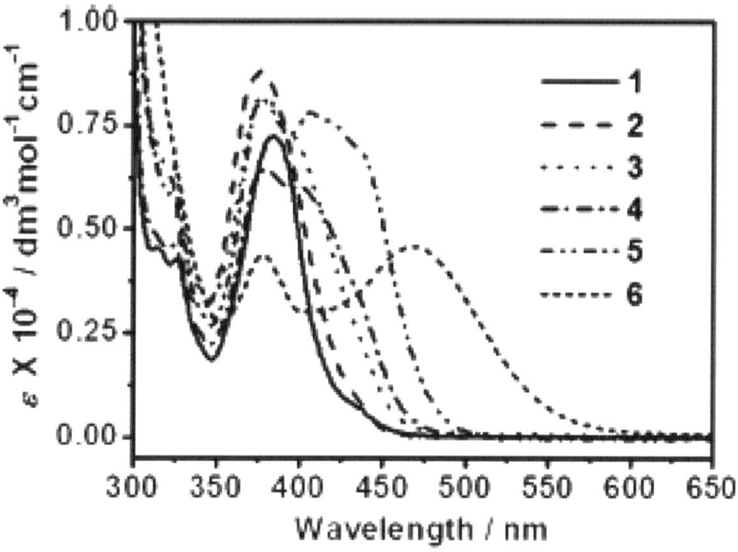

Figure 19.
UV–Vis spectra of pyridine-based N-Heterocyclic Carbene Platinum(II) Complexes 1–6 in ACN at 298 K (Reprinted with permission from Chem. Eur. J. 2013, 19, 10,360–10,369. Copyright@ 2013 Wiley-VCH).

absorption band is redshifted due to increasing the π-electron-donating property of arylalkynyl ligand is assigned as MLCT/ LLCT transition.

The MLCT/LLCT absorption band is sensitive towards polarity of the solvents and shows a blue shifted absorption band from DCM (382 nm) to ACN (376 nm). A negative solvatochromism is observed in Pt (II)–polypyridine [82, 83], and Pt (II)–bzimpy complexes [76, 79] because of decreasing dipole moment during electronic transition.

Non-emissive nature of $[Pt^{II}(C^N^C)Cl][PF_6]$ complex in ACN is can be explained on the basis of low-energy d-d ligand field (LF) states, and effective quenched ^{3}MLCT/^{3}IL state [86]. In contrast to the $[Pt^{II}(C^N^C)Cl][PF_6]$ complex, the tridentate pyridine-based N-heterocyclic carbene ligand based alkynylplatinum(II) complexes 1–5 (**Figure 18**) exhibit strong luminescence in solution with gaussian shaped emission bands (range: 497–631 nm) (**Figure 20**). Only alkynylplatinum(II) complex 6 (**Figure 18**) shows non-emissive character in solution. Interestingly, all alkynylplatinum(II) complexes 1–6 (**Figure 18**) have shown emissive character at low temperature in solid state and glass matrices at 77 K.

The large Stokes shifts and lifetimes in the microsecond are originated from triplet energy state. The emission bands are appeared from an predominantly ^{3}MLCT excited state of [dπ(Pt) → π* (C^N^C)] transition, along with ^{3}LLCT [π(C ≡ CR) → π*(C^N^C)] transition (**Figure 20**). Moreover, CT band and emission band of these metal complexes is altered depending on the nature of the substituted phenyl ring of alkynyl ligands in solution.

The intense luminescence from green to yellow and high PL quantum yield of alkynylplatinum(II) complexes 1–5 can be readily achieved by alternation of alkynyl ligands. The electron-rich moiety quenches the luminescence from ^{3}MLCT excited state via photoinduced electron transfer (PET) process is responsible for non-emissive property of complex 6 in solution [82, 83]. Depending upon increasing the polarity of the solvents, excited state (^{3}MLCT/^{3}LLCT) is lesser stabilized compared to its ground state, leading to a blue shift of absorption spectra in solution, shows negative solvatochromism for alkynylplatinum(II) complex 2. A red shift of the emission band state (559–640 nm) has also been observed for complexes 2–4 in solid at room temperature. The low-energy emission band (559–640 nm) is originated from triplet states (due to presence of metal to-ligand charge transfer (MMLCT) character) of alkynylplatinum(II) complexes 2–4 due to presence of significant contribution from Pt···Pt interaction in solid state [86].

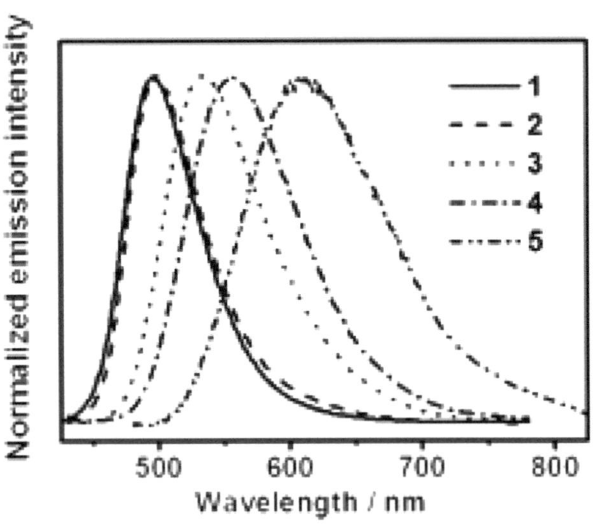

Figure 20.
*Normalized emission spectra of pyridine-based N-Heterocyclic Carbene Platinum(II) Complexes **1–5** in ACN at 298 K (Reprinted with permission from Chem. Eur. J. 2013, 19, 10,360–10,369. Copyright@ 2013 Wiley-VCH).*

4.2 N-heterocyclic carbene Ir (III) complexes and their applications to deep-blue phosphorescent organic light-emitting diodes

Absorption and emission spectra of N-heterocyclic carbene Ir (III) complexes 1–3 (**Figure 21**) are measured in DCM (**Figure 22**). The absorption band at around 320 nm is due to overlap of the $\pi \rightarrow \pi^*$ transition of triazolate chelate, the benzyl (carbene) and pyridyl (triazolate fragment) and considered as LLCT transition. Furthermore, spin-orbit coupling is enhanced by iridium and plays significant role on triplet absorption cross section.

N-heterocyclic carbene Ir (III) complexes 1–3 (**Figure 21**) show emission band at 461, 460, and 458 nm, respectively in DCM. The weak phosphorescence intensity of complex 1 is observed at 392 nm and also indicated by its low quantum efficiency (QE) (only 5.0×10^{-4}). It is observed that the fluorescence quantum yield of complex 2 and 3, is much higher than that of complex 1. The radiative lifetimes of N-heterocyclic carbene Ir (III) complexes 1–3 confirm their phosphorescent nature. The nonradiative decay rate constants (k_{nr}) are found to be 1.2×10^9, 3.5×10^6 and 7.0×10^5 s^{-1} with large differences in quantum efficiency for complex 1–3 respectively.

5. Applications

The carbene chemistry became more popular for their applications as organocatalysts [87, 88] and transition metal catalysts in the synthesis of complex

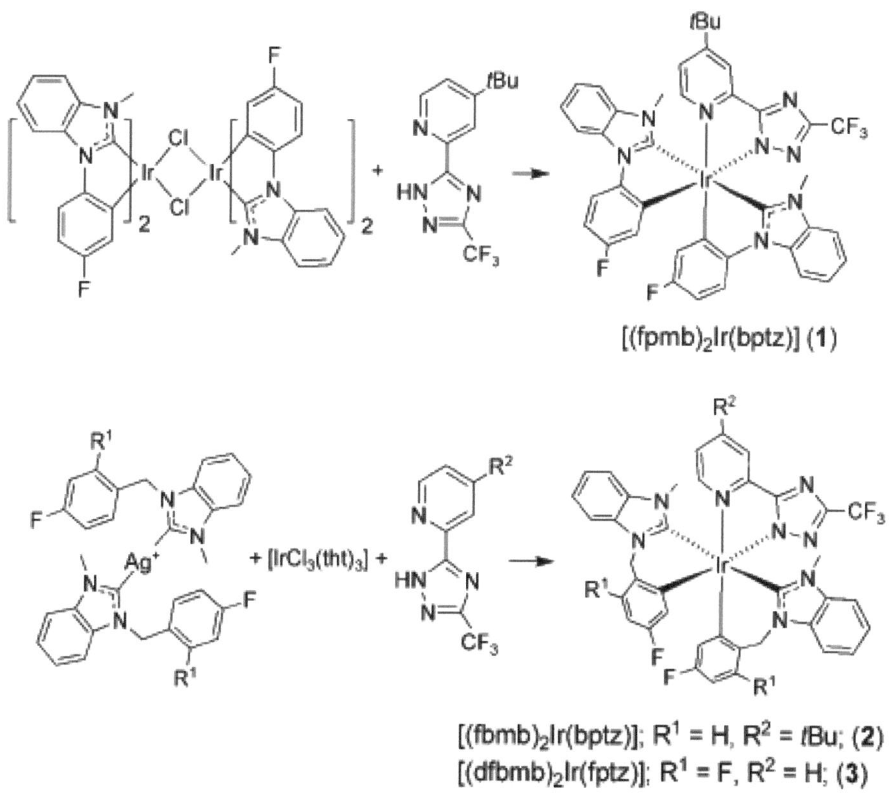

Figure 21.
Structures of N-heterocyclic carbene irridium (III) complexes (1–3).

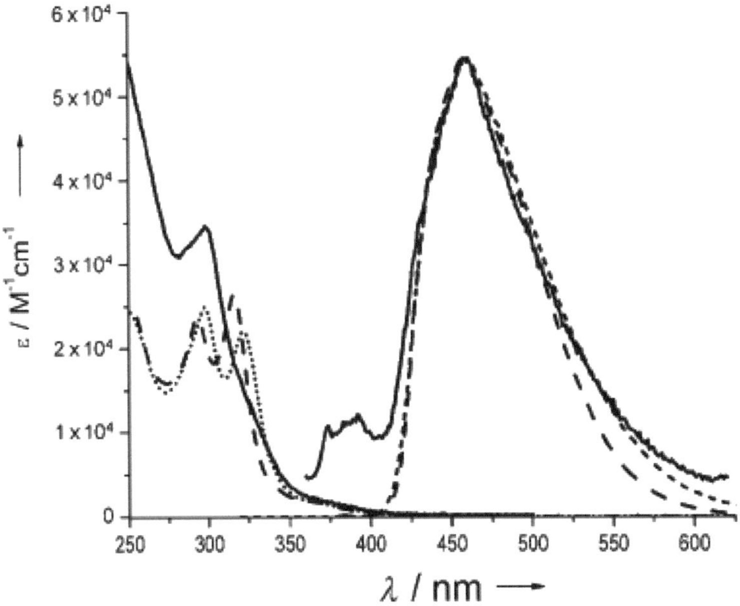

Figure 22.
Absorption and fluorescence spectra of 1 (——), 2 (.....) and 3 (—) in CH_2Cl_2 at 298 K (Reprinted with permission from Angew. Chem. Int. Ed. 2008, 47, 4542. Copyright@ 2008 Wiley-VCH).

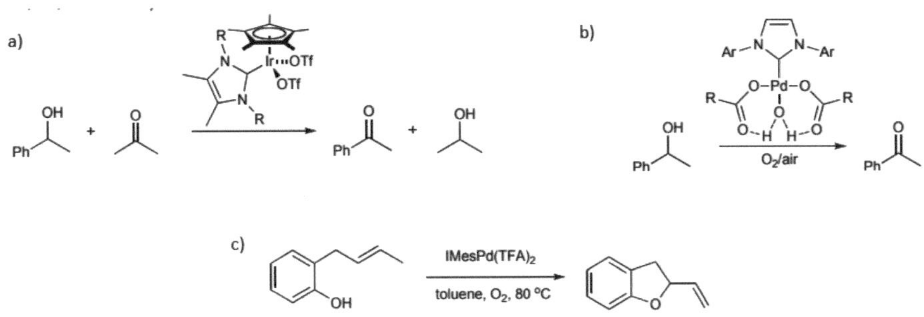

Figure 23.
Types of oxidation reactions catalyzed by NHC-metal complexes.

molecules [89, 90]. The N-heterocyclic carbene (NHCs) is widely used in organometallic chemistry during the last few years [47]. A brief summary of the applications include:

5.1 Oxidation reactions catalyzed by NHC-metal complexes

They can be employed for various types of oxidation reactions like (a) Opppenauer-Type Alcohol Oxidation [91] where smaller R groups show catalytic activity, (b) Palladium-Catalyzed Aerobic Alcohol Oxidation [92] or (c) Wacker-Type Oxidation [93] as shown in **Figure 23**.

In addition, NHC metal complexes have been extensively used for **oxidative cleavage of alkenes** [94] as well as **oxidation of methane** [33]. In this regard, it has been observed that electron deficient alkenes react slower than electron rich alkenes. The NHC-Ru complex remains stable throughout the course of oxidation reaction (**Figure 24**).

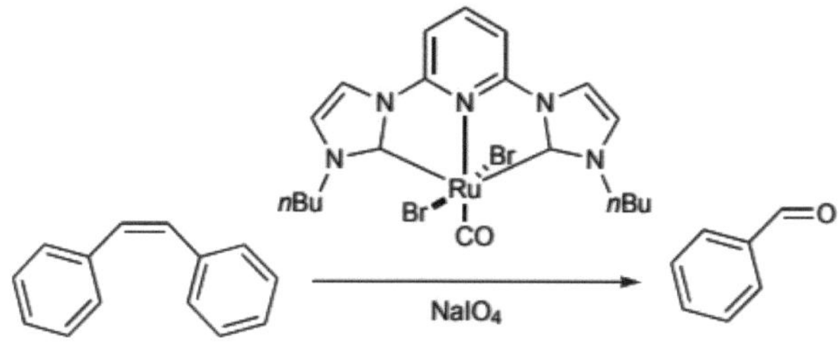

Figure 24.
NHC-Ru complex catalyzed oxidation of alkene.

5.2 Palladium catalyzed reactions

NHCs are also frequently used in **Palladium catalyzed reactions** forming C-C bonds. Mori et al. reported the use of NHC ligand in allylic alkylation with excellent yield [95] (**Figure 25**).

Another important application has been the α-arylation of carbonyl compounds at moderate temperature and short reaction time [96]. The same strategy has been applied for esters and amides [60, 97] (**Figure 26**).

Besides these, NHC act as great ligands for Pd catalyzed various coupling reactions like Heck reaction [98], Negishi reaction [99], Sonagashira reaction [100], Suzuki-Miyaura reaction [101], Stille coupling [102], and Buchwald-Hartwig reaction [103].

The NHC ligands act as good catalysts for tandem coupling reactions as well [104]. The reaction proceeds via amination route (**Figure 27**).

5.3 NHC Complexes in Olefin Metathesis

After the development of Grubbs I catalyst several modifications were carried out to develop more efficient catalysts. In this regard, NHC-Ruthenium complex

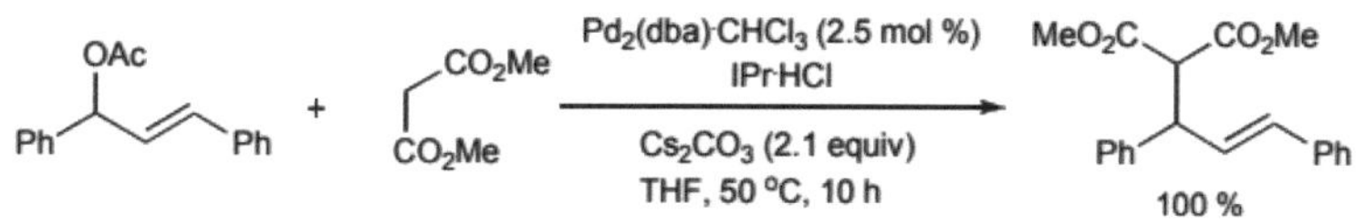

Figure 25.
Use of NHC ligands in allylic alkylation.

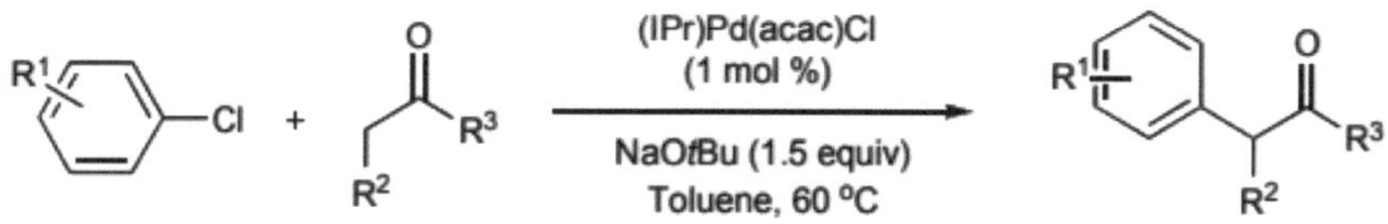

Figure 26.
NHC mediated coupling reaction.

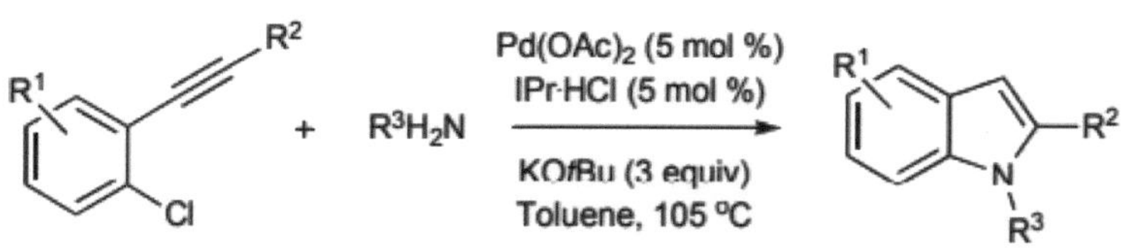

Figure 27.
NHC-metal complex for tandem coupling reactions.

was synthesized for RCM [105]. But the initial design did not show marked difference in activity compared to Grubbs I catalyst, Later on, some combination catalysts were developed which showed greater activity and selectivity [106–109]. Nowadays, these second-generation Grubbs' olefin metathesis catalysts are widely used for metathesis reactions (**Figure 28**).

5.4 NHC as ligands in asymmetric synthesis

There are different approaches for inducing chirality by NHC ligands. One of the ways is N-substituents containing centers of chirality. But initially this method did not show great stereoselectivity [110] but later on the development of bidentate ligands enhanced the enantioselectivity. Grubbs et al. developed another method wherein, NHC ligands had the chiral elements within the N-heterocycles [111]. These ligands showed great selectivity. Another interesting method involves an element of chirality like axial chirality or planer chirality to make the NHC ligands as stereo directing ligands for asymmetric synthesis [112, 113] (**Figure 29**).

5.5 NHC as organocatalysts

Carbene can act as an organocatalyst was demonstrated long back in 1940s [114]. Since then, several attempts have been undertaken to develop NHC mediated Benzoin, Acyloin condensation reactions [115–119] (**Figure 30**).

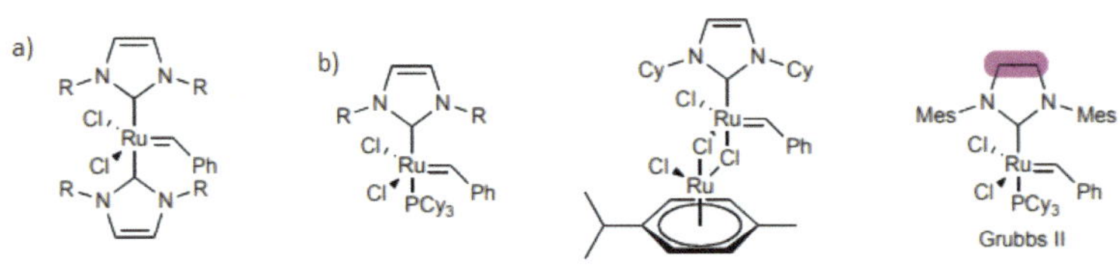

Figure 28.
NHC ligands in metathesis reactions (a) first complex (b) combination catalysts.

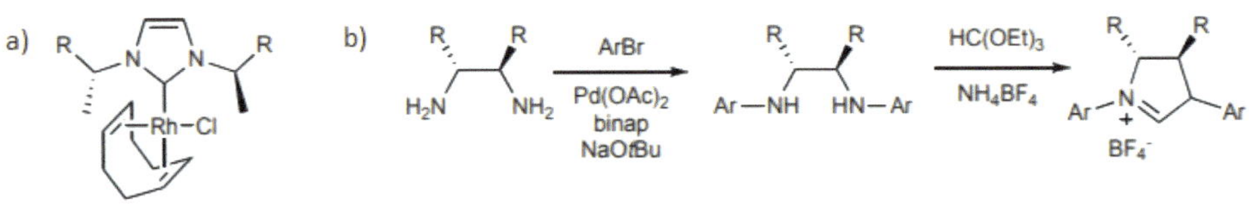

Figure 29.
(a) Chiral NHC ligand structure, (b) synthesis of chiral ligand.

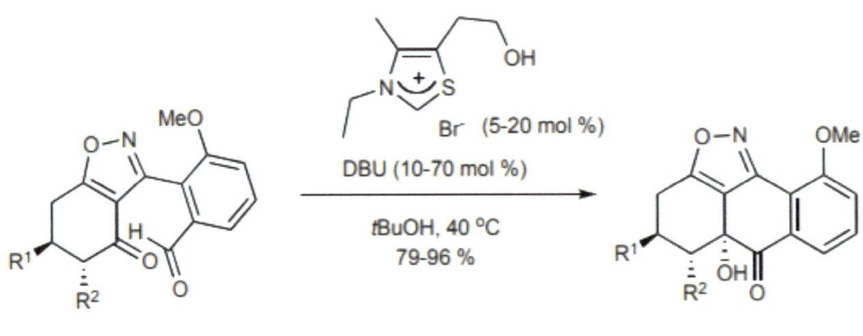

Figure 30.
NHC ligands used as organocatalysts.

6. Conclusion

Carbene has been all along an important reactive intermediate in organic chemistry but the development of N-Heterocyclic carbene has revolutionized the field of organic synthesis. NHC owing to their ease and flexibility of synthesis, their interesting structural properties and in particular their stability have received a great deal of attention within the last few years. There are several strategies for the development of stable and versatile NHC ligands. We have highlighted in this book chapter some general synthetic strategies for the synthesis of some interesting NHC. Along with the discussion on the different strategies adopted in the synthesis of NHC, we focused primarily on some of the salient features of their structures and photophysical properties. Finally, we focus briefly on the various uses of NHC in organic synthesis. Over the years, NHCs have been extensively studied for several applications in organic transformations. Herein, we have very briefly touched upon some of those reactions.

The recent advances in the design of novel ligands, development of interesting structures of NHC enable their exciting potential applications in organic synthesis in the future.

Author details

Arunava Manna[1*], Abhineet Verma[1], Sumit K. Panja[1,2] and Satyen Saha[1*]

1 Department of Chemistry, Institute of Science, Banaras Hindu University, Varanasi, UP, India

2 Tarsadia Institute of Chemical Science, Uka Tarsadia University, Bardoli, Gujrat, India

*Address all correspondence to: arunavam@bhu.ac.in
and satyen.saha@gmail.com

References

[1] Bourissou D, Guerret O, Gabbaï F, Bertrand G. Stable carbenes. Chemical Reviews. 1999;**100**:39-92

[2] Hahn FE, Jahnke MC. Heterocyclic carbenes: Synthesis and coordination chemistry. Angewandte Chemie, International Edition. 2008;**47**: 3122-3172

[3] Despagnet-Ayoub E, Grubbs R. A stable four-membered N-heterocyclic carbene. Journal of the American Chemical Society. 2004;**126**: 10198-10199

[4] Despagnet-Ayoub E, Ritter T. N-heterocyclic carbenes as ligands for olefin metathesis catalysts. N-Heterocyclic Carbenes in Transition Metal Catalysis. 2006;**21**: 193-218

[5] Altenhoff G, Goddard R, Lehmann C, Glorius F. An N-heterocyclic carbene ligand with flexible steric bulk allows suzuki cross-coupling of sterically hindered aryl chlorides at room temperature. Angewandte Chemie, International Edition. 2003;**42**: 3690-3693

[6] Altenhoff G, Goddard R, Lehmann C, Glorius F. Sterically demanding, bioxazoline-derived N-heterocyclic carbene ligands with restricted flexibility for catalysis. Journal of the American Chemical Society. 2004;**126**: 15195-15201

[7] Burstein C, Lehmann C, Glorius F. Imidazo [1,5-A] pyridine-3-ylidenes—pyridine derived N-heterocyclic carbene ligands. Tetrahedron. 2005;**61**: 6207-6217

[8] Alcarazo M, Roseblade S, Cowley A, Fernández R, Brown J, Lassaletta J. Imidazo [1,5-A] pyridine:a versatile architecture for stable N-heterocyclic carbenes. Journal of the American Chemical Society. 2005;**127**:3290-3291

[9] Lavallo V, Canac Y, Präsang C, Donnadieu B, Bertrand G. Stable cyclic (Alkyl) (amino) carbenes as rigid or flexible, bulky, electron-rich ligands for transition-metal catalysts: A quaternary carbon atom makes the difference. Angewandte Chemie, International Edition. 2005;**44**:5705-5709

[10] Lavallo V, Canac Y, DeHope A, Donnadieu B, Bertrand G. A rigid cyclic (Alkyl) (amino) carbene ligand leads to isolation of low-coordinate transition-metal complexes. Angewandte Chemie, International Edition. 2005;**44**: 7236-7239

[11] Yamashita M, Goto K, Kawashima T. Fixation of both O_2 and Co_2 from air by a crystalline palladium complex bearingn-heterocyclic carbene ligands. Journal of the American Chemical Society. 2005;**127**:7294-7295

[12] Raubenheimer H, Lindeque L, Cronje S. Synthesis and characterization of neutral and cationic diamino carbene complexes of gold (I). Journal of Organometallic Chemistry. 1996;**511**: 177-184

[13] Köcher C, Herrmann W. Heterocyclic carbenes. One-pot synthesis of rhodium and iridium carbene complexes. Journal of Organometallic Chemistry. 1997;**532**: 261-265

[14] Hahn F, Wittenbecher L, Boese R, Bläser D. N, N′-Bis(2,2-Dimethylpropyl) Benzimidazolin-2-Ylidene: A stable nucleophilic carbene derived from Benzimidazole. Chemistry—A European Journal. 1999; **5**:1931-1935

[15] Hahn F, Wittenbecher L, Le Van D, Fröhlich R. Evidence for an equilibrium

between an N-heterocyclic carbene and its dimer in solution. Angewandte Chemie, International Edition. 2000;**39**: 541-544

[16] Tan K, Bergman R, Ellman J. Intermediacy of an N-heterocyclic carbene complex in the catalytic C – H activation of a substituted benzimidazole. Journal of the American Chemical Society. 2002;**124**:3202-3203

[17] Metallinos C, Barrett F, Chaytor J, Heska M. A benzimidazole-based N-heterocyclic carbene derived from 1,10-phenanthroline. Organic Letters. 2004; **6**:3641-3644

[18] Hadei N, Kantchev E, O'Brie C, Organ M. Electronic nature of N-heterocyclic carbene ligands: Effect on the suzuki reaction. Organic Letters. 2005;**7**:1991-1994

[19] O'Brien C, Kantchev E, Chass G, Hadei N, Hopkinson A, Organ M, et al. Towards the rational design of palladium-N-heterocyclic carbene catalysts by a combined experimental and computational approach. Tetrahedron. 2005;**61**:9723-9735

[20] Weiss R, Reichel S, Handke M, Hampel F. Generation and trapping reactions of a formal 1:1 complex between singlet carbon and 2,2′-bipyridine. Angewandte Chemie, International Edition. 1998;**37**:344-347

[21] Weiss R, Reichel S. Novel urea derivatives as two-step redox systems. European Journal of Inorganic Chemistry. 2000;**2000**:1935-1939

[22] Nonnenmacher M, Kunz D, Rominger F, Oeser T. First examples of dipyrido [1,2-C:2′,1′-E] imidazolin-7-ylidenes serving as NHC-ligands: Synthesis, properties and structural features of their chromium and tungsten pentacarbonyl complexes. Journal of Organometallic Chemistry. 2005;**690**: 5647-5653

[23] Nonnenmacher M, Kunz D, Rominger F, Oeser T. X-ray crystal structures of 10Π- and 14Π-electron pyrido-annelated N-heterocyclic carbenes. Chemical Communications. 2006;**13**:1378

[24] Alder RW, Blake ME, Bortolotti C, Bufali S, Butts CP, Linehan E, et al. Complexation of stable carbenes with alkali metals. Chemical Communications. 1999;**3**:241-242

[25] Guillen F, Winn C, Alexakis A. Enantioselective copper-catalyzed conjugate addition using chiral diaminocarbene ligands. Tetrahedron: Asymmetry. 2001;**12**:2083-2086

[26] Mayr M, Wurst K, Ongania K, Buchmeiser M. 1,3-dialkyl- and 1,3-diaryl-3,4,5,6-tetrahydropyrimidin-2-ylidene rhodium(I) and palladium(II) complexes: Synthesis, structure, and reactivity. Chemistry—A European Journal. 2004;**10**:1256-1266

[27] Yun J, Marinez E, Grubbs R. A new ruthenium-based olefin metathesis catalyst coordinated with 1,3-dimesityl-1,4,5,6-tetrahydropyrimidin-2-ylidene: Synthesis, X-ray structure, and reactivity. Organometallics. 2004;**23**: 4172-4173

[28] Bazinet P, Yap G, Richeson D. Constructing a stable carbene with a novel topology and electronic framework. Journal of the American Chemical Society. 2003;**125**:13314-13315

[29] Präsang C, Donnadieu B, Bertrand G. Stable planar six-Π-electron six-membered N-heterocyclic carbenes with tunable electronic properties. Journal of the American Chemical Society. 2005;**127**:10182-10183

[30] Scarborough C, Grady M, Guzei I, Gandhi B, Bunel E, Stahl S. Pdii complexes possessing a seven-membered N-heterocyclic carbene ligand. Angewandte Chemie,

International Edition. 2005;**44**: 5269-5272

[31] Kastrup C, Oldfield S, Rzepa H. An Ab initio computational study of monodentate palladium ligand complexes with möbius-aromatic chiral character. Journal of the Chemical Society, Dalton Transactions. 2002;**12**: 2421

[32] Boydston A, Williams K, Bielawski C. A modular approach to main-chain organometallic polymers. Journal of the American Chemical Society. 2005;**127**: 12496-12497

[33] Muehlhofer M, Strassner T, Herrmann W. New catalyst systems for the catalytic conversion of methane into methanol. Angewandte Chemie, International Edition. 2002;**41**: 1745-1747

[34] Fränkel R, Kernbach U, Bakola-Christianopoulou M, Plaia U, Suter M, Ponikwar W, et al. Homoleptic carbene complexes. Journal of Organometallic Chemistry. 2001;**617-618**:530-545

[35] Ofele KJ. A new transition-metal carbon complexe 1, 3-dimethyl –4-imidazolinylid-2-ene pentacarbonylchromium. Journal of Organometallic Chemistry. 1968;**12**: 42-43

[36] Wanzlick H-W, Schönherr H-J. Direct synthesis of a mercury salt-carbene complex. Angewandte Chemie, International Edition. 1968;**7**:141-142

[37] Cardin DJ, Cetinkaya B, Lappert MF, Manojlović-Muir L, Muir KW. An electron-rich olefin as a source of co-ordinated carbene; synthesis of trans-Ptcl2 [C(Nphch2)2] Pet3. Journal of the Chemical Society [Section] D: Chemical Communications. 1971;**5**:400-401

[38] Igau A, Grutzmacher H, Baceiredo A, Bertrand G. Analogous. Alpha., Alpha.'-Bis-Carbenoid, triply bonded species: Synthesis of a stable lambda 3-phosphino carbene-.lambda.5-phosphaacetylene. Journal of the American Chemical Society. 1988;**110**: 6463-6466

[39] Igau A, Baceiredo A, Trinquier G, Bertrand G. [Bis (Diisopropylamino) Phosphino] trimethylsilyl carbene: A stable nucleophilic carbene. Angewandte Chemie, International Edition in English. 1989;**28**: 621-622

[40] Arduengo A, Dias H, Harlow R, Kline M. Electronic stabilization of nucleophilic carbenes. Journal of the American Chemical Society. 1992;**114**: 5530-5534

[41] Schuster O, Yang L, Raubenheimer H, Albrecht M. Beyond conventional N-heterocyclic carbenes: Abnormal, remote, and other classes of NHC ligands with reduced heteroatom stabilization. Chemical Reviews. 2009; **109**:3445-3478

[42] Crabtree R. Abnormal, mesoionic and remote N-heterocyclic carbene complexes. Coordination Chemistry Reviews. 2013;**257**:755-766

[43] Huang D, Zhao P, Astruc D. Catalysis by 1, 2, 3-triazole- and related transition-metal complexes. Coordination Chemistry Reviews. 2014; **272**:145-165

[44] Vivancos Á, Segarra C, Albrecht M. Mesoionic and related less heteroatom-stabilized N-heterocyclic carbene complexes: Synthesis, catalysis, and other applications. Chemical Reviews. 2018;**118**:9493-9586

[45] Arduengo A, Harlow R, Kline M. A stable crystalline carbene. Journal of the American Chemical Society. 1991;**113**: 361-363

[46] Edworthy I, Rodden M, Mungur S, Davis K, Blake A, Wilson C, et al. Silver

alkoxide and amino N-heterocyclic carbenes; syntheses and crystal structures. Journal of Organometallic Chemistry. 2005;**690**:5710-5719

[47] Nolan SP, editor. N-heterocyclic carbenes in synthesis. Weinheim, Germany: Wiley-VCH; 2006

[48] Ishii T, Kakeno Y, Nagao K, Ohmiya H. N-heterocyclic carbene-catalyzed decarboxylative alkylation of aldehydes. Journal of American Chemical Society. 2019;**141**(9):3854-3858

[49] Bellotti P, Koy M, Hopkinson MN. Recent advances in the chemistry and applications of N-heterocyclic carbenes. Nature Review Chemistry. 2021;**5**: 711-725

[50] Kamijo S, Yamamoto Y. Recent progress in the catalytic synthesis of imidazoles. Chemistry: An Asian Journal. 2007;**2**:568-578

[51] Boulton AJ, Katritzky AR. Advances in Heterocyclic Chemistry. Vol. 27. New York: Academic Press; 1980. p. 241

[52] Neier R, Grimmett MR, Bellus D, Houben J, Weyl T. Science of Synthesis. Stuttgart: Thieme; 2002. p. 325

[53] Grimmett MR. Comprehensive heterocyclic chemistry: The structure, reactions, synthesis and uses of heterocyclic compounds. In: Katritzky AR, Rees CW, editors. Vol. 5. Oxford: Pergamon; 1984. p. 345

[54] Jahnke MC, Hahn FE. Introduction to N-heterocyclic carbenes: Synthesis and stereoelectronic parameters in N-heterocyclic carbenes: From laboratory curiosities to efficient synthetic tools. Catalysis Series. 2016;**2**:1-45

[55] Leuthäußer S, Schwarz D, Plenio H. Tuning the electronic properties of N-heterocyclic carbenes. Chemistry—A European Journal. 2007;**13**:7195

[56] Rosadiuk K, Bohle D. Anhydrous dinitrogen trioxide solutions for brønsted acid free nitrous acid chemistry. European Journal of Inorganic Chemistry. 2017;**46**:5461-5465

[57] Fleckenstein C, Roy S, Leuthäußer S, Plenio H. Sulfonated N-heterocyclic carbenes for suzuki coupling in water. Chemical Communication. 2007;**27**: 2870-2872

[58] Kantchev E, Ying J. Practical one-pot, three-component synthesis of n-heterocyclic carbene (NHC) ligated palladacycles derived from N,N-dimethylbenzylamine. Organometallics. 2008;**28**(1):289-299

[59] Fliedel C, Maisse-François A, Bellemin-Laponnaz S. Palladium(II) complexes of a Bis-2-aminobiphenyl N-heterocyclic carbene: Synthesis, structural studies and catalytic activity. Inorganica Chimica Acta. 2007;**360**(1): 143-148

[60] Lee S, Hartwig J. Improved catalysts for the palladium-catalyzed synthesis of oxindoles by amide A-arylation. rate acceleration, use of aryl chloride substrates, and a new carbene ligand for asymmetric transformations. The Journal of Organic Chemistry. 2001; **66**(10):3402-3415

[61] Jia Y, Hillgren J, Watson E, Marsden S, Kündig E. Chiral N-Heterocyclic carbene ligands for asymmetric catalytic oxindole synthesis. Chemical Communications. 2008;**34**:4040-4042

[62] Kündig E, Seidel T, Jia Y, Bernardinelli G. Bulky chiral carbene ligands and their application in the palladium-catalyzed asymmetric intramolecular A-arylation of amides. Angewandte Chemie International Edition. 2007;**46**(44):8484-8487

[63] Sauvage X, Demonceau A, Delaude L. Homobimetallic Ruthenium–N-Heterocyclic Carbene Complexes for

Olefin Metathesis. Metathesis Chemistry. 2007;**243**:91-109

[64] Delaude L, Szypa M, Demonceau A, Noels AF. In situ generated ruthenium catalysts bearing N-heterocyclic carbene ligands for the ring-opening metathesis polymerization of cyclooctene. Advanced Synthesis and Catalysis. 2002;**344**:749-756

[65] Waltman A, Grubbs R. A new class of chelating N-heterocyclic carbene ligands and their complexes with palladium. Organometallics. 2004; **23**(13):3105-3107

[66] Vougioukalakis GC, Grubbs RH. Ruthenium olefin metathesis catalysts bearing an N-Fluorophenyl-N-Mesityl-substituted unsymmetrical N-Heterocyclic carbene. Organometallics. 2007;**26**(9): 2469-2472

[67] Ritter T, Day M, Grubbs R. Rate acceleration in olefin metathesis through a fluorine–Ruthenium interaction. Journal of the American Chemical Society. 2006;**128**(36): 11768-11769

[68] Winkelmann O, Linder D, Lacour J, Näther C, Lüning U. Chiral concave imidazolinium salts as precursors to chiral concaven-heterocyclic carbenes. European Journal of Organic Chemistry. 2007;**22**:3687-3697

[69] Xu G, Gilbertson S. Development of building blocks for the synthesis of N-Heterocyclic carbene ligands. Organic Letters. 2005;**7**(21):4605-4608

[70] Makino T, Masu H, Katagiri K, Yamasaki R, Azumaya I, Saito S. Synthesis, structure, and solvent-induced spontaneous homochiral assembly of bidentate Bis(N,N′-Diaryl-N-Heterocyclic Carbene)-palladium complexes. European Journal of Inorganic Chemistry. 2008;**31**: 4861-4865

[71] Kuriyama M, Shimazawa R, Shirai R. Design and synthesis of thioether-imidazolium chlorides as efficient ligands for palladium-catalyzed suzuki–miyaura coupling of aryl bromides with arylboronic acids. Tetrahedron. 2007; **63**(38):9393-9400

[72] Clavier H, Coutable L, Guillemin JC, Mauduit M. New bidentate alkoxy-NHC ligands for enantioselective copper-catalysed conjugate addition. Tetrahedron: Asymmetry. 2005;**16**(5): 921-924

[73] Bildstein B, Malaun M, Kopacka H, Wurst K, Mitterb€ock M, Ongania K-H, et al. N,N'-Diferrocenyl-N-Heterocyclic carbenes and their derivatives. Organometallics. 1999;**18**(21):4325-4336

[74] Van Ausdall B, Glass J, Wiggins K, Aarif A, Louie J. A systematic investigation of factors influencing the decarboxylation of imidazolium carboxylates. The Journal of Organic Chemistry. 2009;**74**(20):7935-7942

[75] Bailey J, Hill M, Marsh R, Miskowski V, Schaefer W, Gray H. Electronic spectroscopy of chloro(terpyridine) platinum(II). Inorganic Chemistry. 1995;**34**:4591-4599

[76] Tam AY-Y, Lam WH, Wong KM-C, Zhu N, Yam VW-W. Luminescent alkynylplatinum(II) complexes of 2,6-Bis(N-Alkylbenzimidazol-2′-Yl)pyridine-type ligands with ready tunability of the nature of the emissive states by solvent and electronic property modulation. Chemistry—A European Journal. 2008;**14**:4562-4576

[77] Chan C-W, Cheng L-K, Che CM. Luminescent donor-acceptor platinum (Ii) complexes. Coordination Chemistry Reviews. 1994;**132**:87-97

[78] Lu W, Chui S, Ng K, Che C. A submicrometer wire-to-wheel metamorphism of hybrid tridentate cyclometalated platinum(II) complexes.

Angewandte Chemie. 2008;**120**: 4644-4648

[79] Tam AY-Y, Wong KM, Yam VW-W. Unusual luminescence enhancement of metallogels of alkynylplatinum(II) 2, 6-bis(N-alkylbenzimidazol-2′-Yl) pyridine complexes upon a gel-to-sol phase transition at elevated temperatures. Journal of the American Chemical Society. 2009;**131**:6253-6260

[80] Arduengo A, Dias H, Calabrese J, Davidson F. Homoleptic carbene-silver(I) and carbene-copper(I) complexes. Organometallics. 1993; **12**(9):3405-3409

[81] Lin I, Vasam C. Preparation and application of N-heterocyclic carbene complexes of Ag(I). Coordination Chemistry Reviews. 2007;**251**(5-6): 642-670

[82] Wong KM, Tang W-S, Lu X-X, Zhu N, Yam VW-W. Functionalized platinum(II) terpyridyl alkynyl complexes as colorimetric and luminescence Ph sensors. Inorganic Chemistry. 2005;**44**:1492-1498

[83] Leung SY-L, Tam AY-Y, Tao C-H, Chow HS, Yam VW-W. Single-turn helix–Coil strands stabilized by metal···metal and Π–Π interactions of the alkynylplatinum(II) terpyridyl moieties in meta-phenylene ethynylene foldamers. Journal of the American Chemical Society. 2011;**134**:1047-1056

[84] Yersin H, Donges D. Low-lying electronic states and photophysical properties of organometallic Pd(II) and Pt(II) compounds. Modern research trends presented in detailed case studies. In: Yersin H, editor. Transition Metal and Rare Earth Compounds. Excited States, Transitions and Interactions. Vol. II. Berlin: Springer-Verlag; 2001. p. 81

[85] Herber RH, Croft M, Coyer MJ, Bilash B, Sahiner A. Origin of polychromism of cis square-planar platinum(II) complexes: Comparison of two forms of [Pt (2, 2'-Bpy)(Cl2)]. Inorganic Chemistry. 1994;**33**:2422-2426

[86] Lee C, Sabiah S, Wang J, Hwang W, Lin I. Water-induced changes of photoluminescence of a pincer-type n-heterocyclic carbene platinum(II) complex. Organometallics. 2010;**29**: 286-289

[87] Enders D, Niemeier O, Henseler A. Organocatalysis by N-heterocyclic carbenes. Chemical Reviews. 2007;**107**: 5606-5655

[88] Flanigan DM, Romanov-Michailidis F, White NA, Rovis T. Organocatalytic reactions enabled by N-heterocyclic carbenes. Chemical Reviews. 2015;**115**: 9307-9387

[89] Díez-González S, Marion N, Nolan S. N-heterocyclic carbenes in late transition metal catalysis. Chemical Reviews. 2009;**109**:3612-3676

[90] Hahn FE. Introduction: Carbene chemistry. Chemical Reviews. 2018;**118**: 9455-9456

[91] Hanasaka F, Fujita K, Yamaguchi R. Cp*Ir complexes bearing N-heterocyclic carbene ligands as effective catalysts for oppenauer-type oxidation of alcohols. Organometallics. 2004;**23**(7):1490-1492

[92] Schultz M, Hamilton S, Jensen D, Sigman M. Development and comparison of the substrate scope of Pd-catalysts for the aerobic oxidation of alcohols. The Journal of Organic Chemistry. 2004;**70**(9):3343-3352

[93] Muñiz K. Palladium-carbene catalysts for aerobic, intramolecular wacker-type cyclisation reactions. Advanced Synthesis and Catalysis. 2004;**346**(12):1425-1428

[94] Poyatos M, Mata J, Falomir E, Crabtree R, Peris E. New ruthenium(II)

CNC-pincer bis(carbene) complexes: Synthesis and catalytic activity. Organometallics. 2003;**22**(5):1110-1114

[95] Sato Y, Yoshino T, Mori M. Pd-catalyzed allylic substitution using nucleophilic N-heterocyclic carbene as a ligand. Organic Letters. 2002;**5**(1):31-33

[96] Navarro O, Marion N, Scott N, González J, Amoroso D, Bell A, et al. Synthesis of novel (NHC)Pd (Acac)Cl complexes (acac = acetylacetonate) and their activity in cross-coupling reactions. Tetrahedron. 2005;**61**(41):9716-9722

[97] Lee S, Beare N, Hartwig J. Palladium-catalyzed A-arylation of esters and protected amino acids. Journal of the American Chemical Society. 2001;**123**(34):8410-8411

[98] Herrmann W, Elison M, Fischer J, Köcher C, Artus G. Metal complexes of N-heterocyclic carbenes—A new structural principle for catalysts in homogeneous catalysis. Angwandte Chemie International Edition. 1995; **34**(21):2371-2374

[99] Hadei N, Kantchev E, O'Brie C, Organ M. The first negishi cross-coupling reaction of two alkyl centers utilizing a Pd—N-heterocyclic carbene (NHC) catalyst. Organic Letters. 2005; **7**(17):3805-3807

[100] Batey R, Shen M, Lough A. Carbamoyl-substituted N-heterocyclic carbene complexes of palladium(II): Application to sonogashira cross-coupling reactions. Organic Letters. 2002;**4**(9):1411-1414

[101] Walker S, Barder T, Martinelli J, Buchwald S. A rationally designed universal catalyst for Suzuki–Miyaura coupling processes. Angewandte Chemie, International Edition. 2004; **43**(14):1871-1876

[102] Weskamp T, Böhm V, Herrmann W. Combining N-heterocyclic carbenes and phosphines: Improved palladium(II) catalysts for aryl coupling reactions. Journal of Organometallic Chemistry. 1999;**585**(2): 348-352

[103] Viciu M, Kelly R, Stevens E, Naud F, Studer M, Nolan S. Synthesis, characterization, and catalytic activity of N-heterocyclic carbene (NHC) palladacycle complexes. Organic Letters. 2003;**5**(9):1479-1482

[104] Ackermann L. General and efficient indole syntheses based on catalytic amination reactions. Organic Letters. 2005;**7**(3):439-442

[105] Herrmann W. N-heterocyclic carbenes: A new concept in organometallic catalysis. Angewandte Chemie, International Edition. 2002; **41**(8):1290-1309

[106] Huang J, Stevens E, Nolan S, Petersen J. Olefin metathesis-active ruthenium complexes bearing a nucleophilic carbene ligand. Journal of the American Chemical Society. 1999; **121**(12):2674-2678

[107] Weskamp T, Kohl F, Hieringer W, Gleich D, Herrmann W. Highly active ruthenium catalysts for olefin metathesis: The synergy of N-heterocyclic carbenes and coordinatively labile ligands. Angewandte Chemie, International Edition. 1999;**38**(16):2416-2419

[108] Ackermann L, Fürstner A, Weskamp T, Kohl F, Herrmann W. Ruthenium carbene complexes with imidazolin-2-ylidene ligands allow the formation of tetrasubstituted cycloalkenes by RCM. Tetrahedron Letters. 1999;**40**(26):4787-4790

[109] Scholl M, Ding S, Lee C, Grubbs R. Synthesis and activity of a new generation of ruthenium-based olefin metathesis catalysts coordinated with

1,3-dimesityl-4,5-dihydroimidazol-2-ylidene ligands. Organic Letters. 1999; **1**(6):953-956

[110] Herrmann W, Goossen L, Köcher C, Artus G. Chiral heterocylic carbenes in asymmetric homogeneous catalysis. Angewandte Chemie, International Edition. 1996;**35**(2324):2805-2807

[111] Seiders T, Ward D, Grubbs R. Enantioselective ruthenium-catalyzed ring-closing metathesis. Organic Letters. 2001;**3**(20):3225-3228

[112] Clyne D, Jin J, Genest E, Gallucci J, RajanBabu T. First chelated chiral N-heterocyclic bis-carbene complexes. Organic Letters. 2000;**2**(8):1125-1128

[113] Broggini D, Togni A. Synthesis and structure of an enantiomerically pure C2 symmetric ferrocenyl carbene. Helvetica Chimica Acta. 2002;**85**(8): 2518-2522

[114] van den Berg HJJ. Molecular Catalysis. 1943;**51**:1-12

[115] Sheehan J, Hunneman D. Homogeneous asymmetric catalysis. Journal of the American Chemical Society. 1966;**88**(15):3666-3667

[116] Ramanjaneyulu BT, Mahesh S, Anand RV. N-Heterocyclic Carbene Catalyzed Highly Chemoselective Intermolecular Crossed Acyloin Condensation of Aromatic Aldehydes with Trifluoroacetaldehyde Ethyl Hemiacetal. Organic Letters. 2015;**17**(1):6-9

[117] Tagaki W, Tamura Y, Yano Y. Asymmetric benzoin condensation catalyzed by optically active thiazolium salts in micellar two-phase media. Bulletin of the Chemical Society of Japan. 1980;**53**(2):478-480

[118] Knight R, Leeper F. Synthesis of and asymmetric induction by chiral bicyclic thiazolium salts. Tetrahedron Letters. 1997;**38**(20):3611-3614

[119] Dvorak CA, Rawal VH. Catalysis of benzoin condensation by conformationally-restricted chiral bicyclic thiazolium salts. Tetrahedron Letters. 1998;**39**(19):2925-2928

Chapter 2

Recent Advances in Cooperative N-Heterocyclic Carbene Catalysis

Hideto Miyabe

Abstract

The N-heterocyclic carbenes (NHCs) open the new field of organocatalysis, leading to the dramatic progress on the cooperative NHC catalysis with transition-metal catalysts or photocatalysts.

Keywords: NHC, catalysis, organocatalyst, transition-metal, photocatalyst

1. Introduction

In the past few decades, N-heterocyclic carbenes (NHCs) open the new field of organocatalysis in synthetic organic chemistry [1]. Particularly, chiral NHCs have gained increasing attention as a powerful and versatile organocatalyst for the enantioselective synthesis of various molecules with structural diversity and complexity [2–7]. Although a wide variety of chiral NHC precursors were developed, chiral thiazolium-derived carbenes are the most widely used catalysts for asymmetric synthesis. Based on the characteristic structures, chiral thiazolium precursors can be classified as aminoindanol-based thiazoliums, morpholine-based thiazoliums, pyrrolidine-based thiazoliums, and acyclic thiazoliums [6]. Furthermore, new methods and strategies for NHC catalysis are emerging continuously, leading to remarkable progress on cooperative catalysis using NHC/Lewis acid, NHC/Brønsted acid, and NHC/hydrogen-bonding organocatalyst [8, 9]. In recent years, cooperative catalysis has been expanded by the combination of NHCs with transition-metal catalysts or photocatalysts [10]. This chapter highlights the recent dramatic progress in the cooperative NHC catalysis with transition-metal catalysts or photocatalysts.

2. Cooperative NHC catalysis with transition-metal catalysts

In recent years, the use of transition-metal catalysts in the NHC catalysis has become a widespread strategy for cooperative catalysis, although NHCs are known to act as a ligand for transition metals.

The palladium-catalyzed allylic substitutions are wildly used for achieving cooperative NHC catalysis. Initially, the successful combination of NHC catalysis with transition-metal catalysis was reported in the cascade reactions involving the addition of NHC-catalyzed product to π-allyl palladium intermediate [11–13]. In 2014, cooperative catalysis was achieved by the simultaneous activation of substrates using NHC catalyst and palladium catalyst [14]. This cooperative transformation proceeded *via* the addition of the Breslow intermediate, generated from the NHC catalyst, into the π-allyl palladium intermediate.

The palladium-catalyzed allylic substitutions are applied to the enantioselective NHC catalysis [15–20]. The cooperative catalysis was achieved by using chiral NHC catalyst and palladium catalyst (**Figure 1**) [15–17]. In the presence of palladium catalyst [$Pd(PPh_3)_4$ (5 mol%)] and chiral NHC generated from NHC precursor **(5a*S*,10b*R*)-A1** (15 mol%) and Cs_2CO_3 (1 equiv), the enantioselective [4 + 3] annulation reaction between vinyl benzoxazinanone **1** and cinnamaldehyde **2** were performed in THF at room temperature. The benzazepine derivative **3** was obtained in 86% yield with 99% ee [15]. The proposed catalytic cycle involves the NHC-catalyzed activation of enal **2** followed by the Pd(0)-catalyzed allylic alkylation. Initially, the palladium-catalyzed decarboxylation of vinyl benzoxazinanone **1** gives the π-allyl palladium(II) complex, which reacts with the azolium homoenolate generated from cinnamaldehyde **2** and NHC. The subsequent cyclization provides benzazepine **3** accompanied by the regeneration of the NHC catalyst. In this communication, the stereochemical outcome was explained by the proposed transition state, in which the formation of

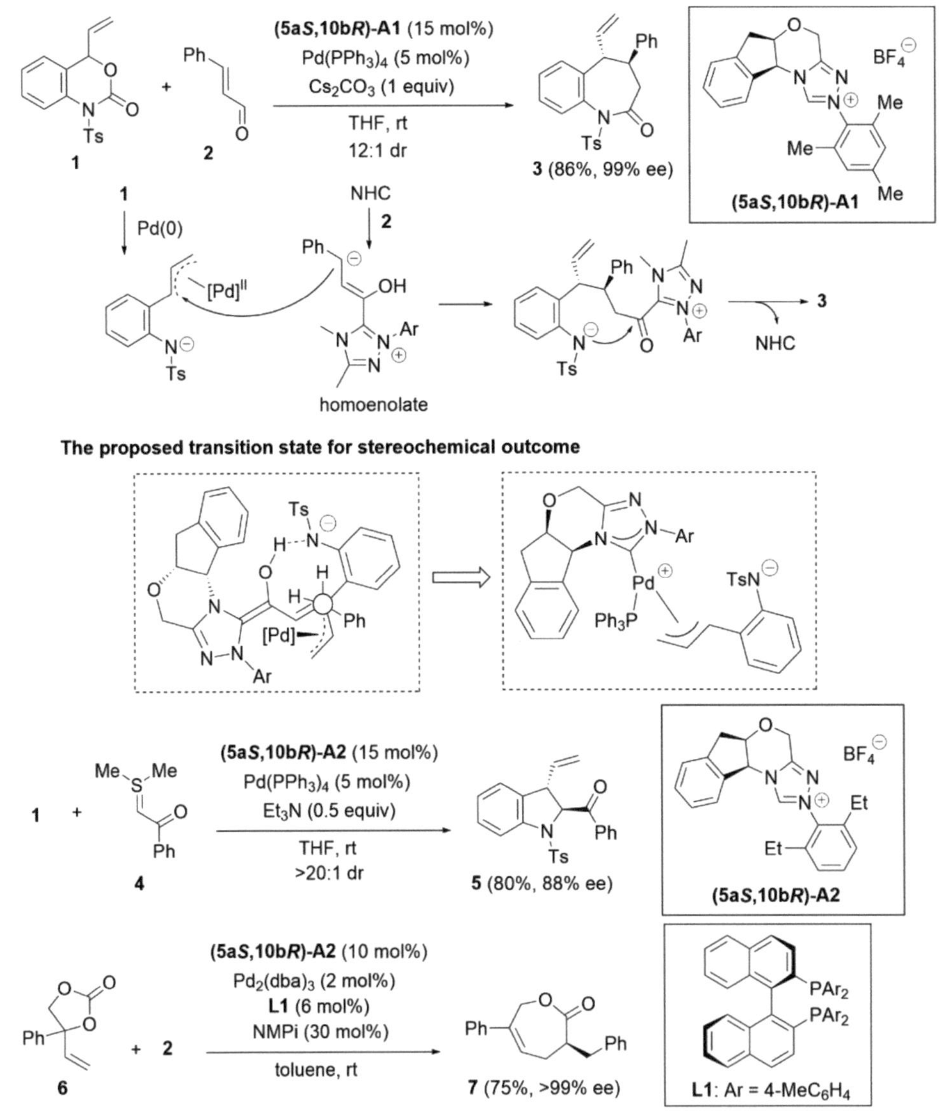

Figure 1.
Enantioselective catalysis using NHC and π-allyl palladium(II) complex.

hydrogen-bonding interaction promotes allylic substitution. Later, a comprehensive investigation of the mechanism was conducted to understand the features of this reaction [16]. A detailed study shows that NHC not only serves as an organocatalyst to activate enal **2** but also a ligand of palladium. Furthermore, the cooperative catalysis was applied to the enantioselective [4 + 1] annulation between benzoxazinanone **1** and sulfur ylide **4**. When NHC precursor **(5a*S*,10b*R*)-A2** and $Pd(PPh_3)_4$ were employed, the desired annulation product **5** was obtained in 80% yield with 88% ee [16]. The combination of chiral NHC, generated from precursor **(5a*S*,10b*R*)-A2**, and a chiral palladium catalyst, generated from $Pd_2(dba)_3$ and ligand **L1**, promoted the highly enantioselective [5 + 2] annulation reaction between phenyl vinylethylene carbonate **6** and cinnamaldehyde **2** [17]. In this reaction, the use of a bidentate phosphine ligand **L1** is crucial to prevent the coordination of NHC to the active Pd catalyst.

NHCs can invert the reactivity of aldehyde from electrophilic to nucleophilic by the formation of Breslow intermediate as an acyl anion equivalent from NHC catalyst and aldehyde. The cooperative NHC/palladium reactions through the nucleophilic addition of Breslow intermediate to the π-allyl palladium(II) complex were investigated (**Figure 2**) [21–26]. The 2:1 coupling reaction of pyridine-2-carboxaldehyde **8** and allyl acetate **9** has been developed [21]. Under the optimized reaction conditions using $Pd(PPh_3)_4$ and NHC generated from precursor **A3** and triethylamine, 2-methyl-1,4-di(pyridin-2-yl)butane-1,4-dione **10** was obtained in 83% yield as a 2:1 coupling product. The proposed catalytic cycle involves the formation of Breslow intermediate as an acyl anion equivalent from NHC catalyst and aldehyde **8** through the addition of NHC to the formyl group of **8** followed by the proton transfer. Next, the addition of Breslow intermediate to the π-allyl palladium(II) complex, generated from allyl acetate **9** and $Pd(PPh_3)_4$, leads to the formation of unsaturated ketone **11** *via* the liberation of NHC. In this transformation, the N atom of the pyridine ring acts as a coordination site toward the palladium of the π-allyl complex. Finally, ketone **11** is converted to product **10** through condensation with another Breslow intermediate. This cooperative catalysis was extended to *C*-glycosylation using aldehyde **8** and glucal **12** [22].

The propargylation reaction of pyridine-2-carboxaldehyde **8** was also developed [23]. The propargylic ketone product **15** was obtained in 74% when propargylic carbonate **14** was used under the cooperative NHC/palladium catalysis conditions. Furthermore, the reaction of widely available aldehydes with diarylmethyl carbonates was studied [24]. When aliphatic aldehyde **16** and diarylmethyl carbonates **17** were employed under the cooperative conditions using NHC precursor **A4**, α-arylated ketone **18** was obtained in 78% yield.

The cooperative NHC/palladium reaction for the umpolung 1,4-addition of aryl iodides or vinyl bromides to enals was developed [27, 28]. The combination of NHC, generated from precursor **A5**, and a palladium catalyst, generated from $Pd_2(dba)_3$ and ligand **L3**, promoted the 1,4-addition of iodobenzene **19** to cinnamaldehyde **2** to give methyl β,β-diphenyl propanoate **20** in 71% yield (**Figure 3**) [27]. This reaction is the palladium-catalyzed arylation of NHC-bound homoenolate equivalent generated from cinnamaldehyde **2** and NHC. The oxidative addition of palladium catalyst to iodobenzene **19** generates the activated $PhPdI(L_n)$ as an electrophile, which reacts with nucleophilic homoenolate equivalent. The subsequent reductive elimination provides the NHC-bound intermediate, which reacts with MeOH to afford methyl β,β-diphenyl propanoate **20**. Additionally, 1,4-addition of vinyl bromides to enals was also studied under similar reaction conditions [28].

The cooperative catalysis using NHC and copper catalyst was investigated (**Figure 4**) [29, 30]. The catalytic reaction using alkyne **21**, tosyl azide **22**, and isatin-derived imine **23** was investigated by using NHC precursor **A6** and CuI [29]. The spiro-azetidine oxindole **24** was obtained in 83% yield with 85:15 er. Initially,

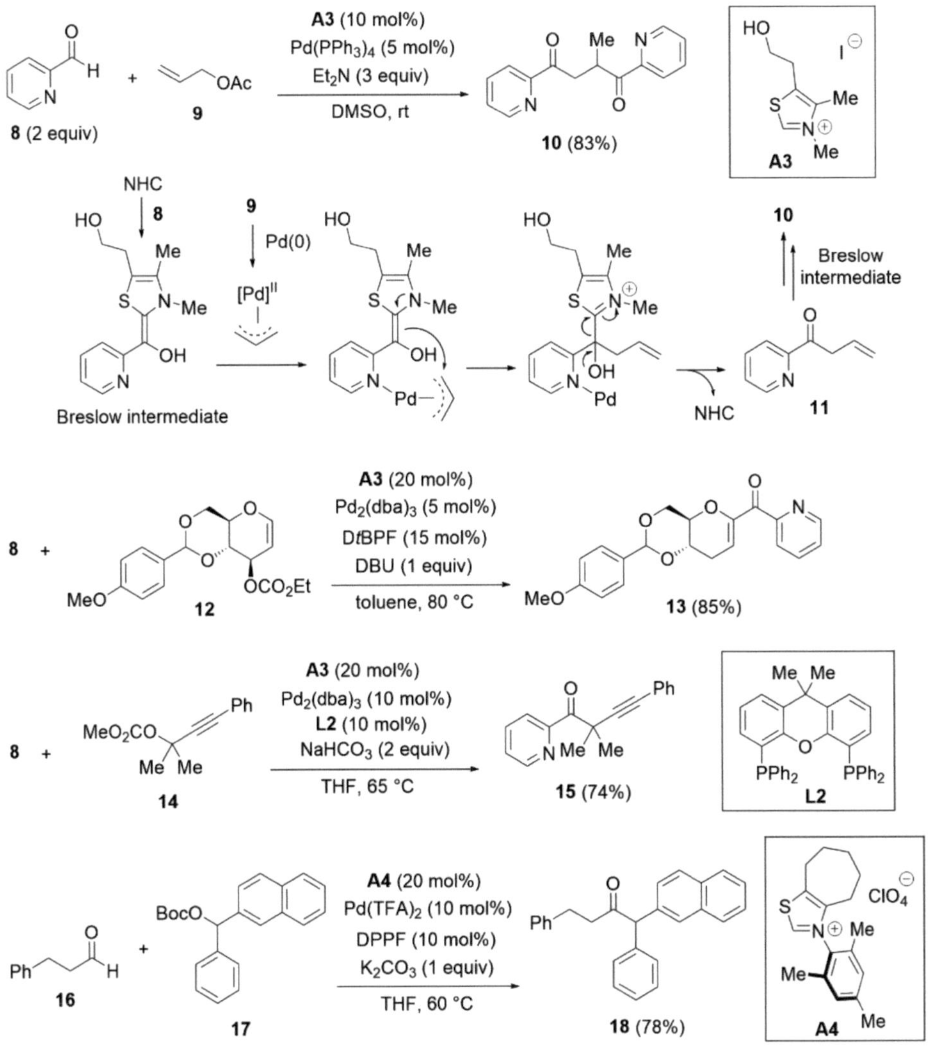

Figure 2.
Cooperative catalysis based on umpolung of aldehydes.

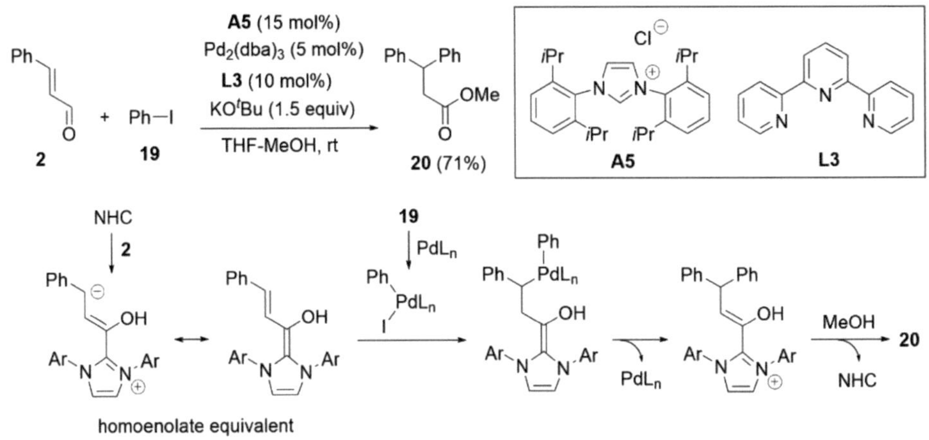

Figure 3.
Cooperative catalysis for umpolung 1,4-addition to cinnamaldehyde.

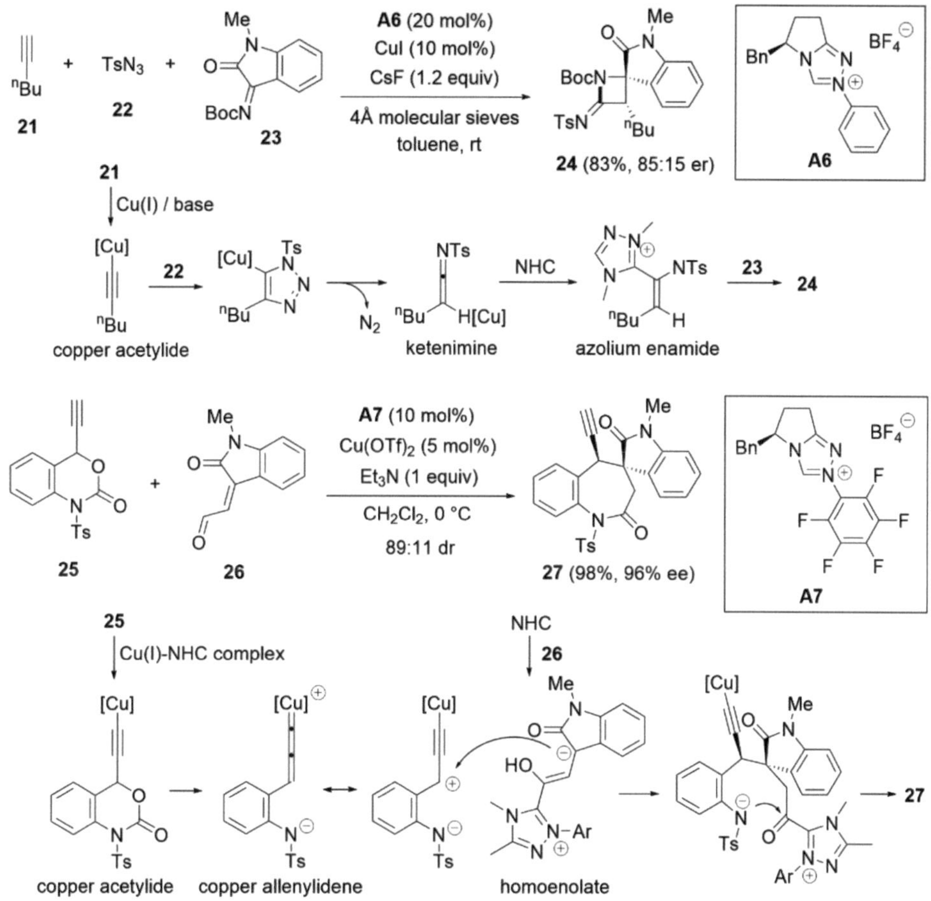

Figure 4.
Cooperative catalysis with copper catalyst.

copper acetylide is generated from **21** and Cu(I) under the basic conditions. The proposed catalytic cycle involves the formation of ketenimine intermediate *via* triazole generated by [3+2] cycloadditions between copper acetylide and azide **22**. Subsequently, ketenimine reacts with NHC to form azolium enamide, which undergoes the formal [2+2] cycloaddtion with imine **23** to afford product **24**. In the presence of NHC precursor **A7** (10 mol%), $Cu(OTf)_2$ (5 mol%) and Et_3N (1 equiv), [4 + 3] annulation between ethynyl benzoxazinanone **25** and isatin-derived enal **26** led to the formation of spirooxindole **27** in 98% yield with 96% ee [30]. In this catalysis, the decarboxylation of copper acetylide leads to copper allenylidene, which reacts with the NHC-linked homoenolate generated from enal **26**. Since NHC serves as a ligand of copper, chiral Cu(I)-NHC complex would participate in the control of stereochemistry, together with chiral NHC catalyst.

The cooperative catalysis using NHC and gold catalyst was reported (**Figure 5**) [31]. When NHC precursor **A8** (20 mol%), $PPh_3AuCl/AgPF_6$ (10 mol%), and DABCO (25 mol%) were employed, the relay reaction of ynamide substrate **28** with enal **29** gave the bicyclic lactam **30** in 84% yield with 99% ee. In this catalysis, Au(I) promotes the cyclization of ynamide **28** to generate α,β-unsaturated imine as a key intermediate. The subsequent reaction of α,β-unsaturated imine with the NHC-linked enolate, generated from enal **29** and NHC, gives the bicyclic product **30**.

The combination of NHC catalysis and ruthenium redox catalysis was investigated [32–34]. The oxidation of the Breslow intermediate leads to the formation of

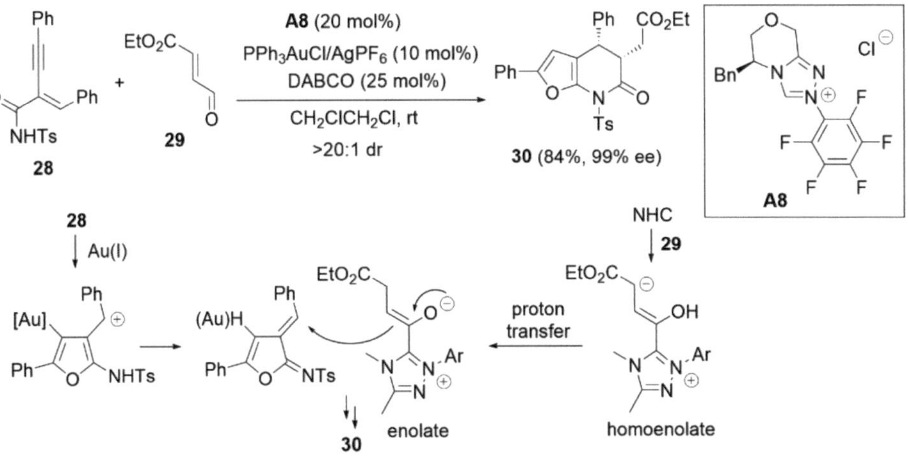

Figure 5.
Cooperative catalysis with gold catalyst.

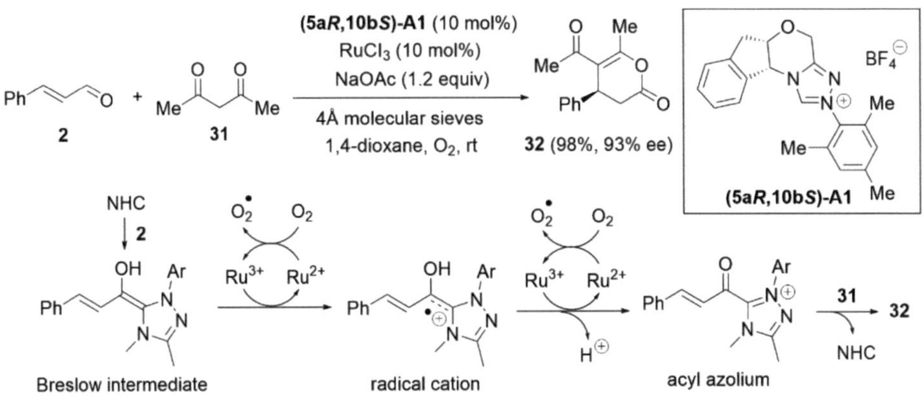

Figure 6.
Cooperative catalysis with ruthenium catalyst.

α,β-unsaturated acyl azolium *via* radical cation (**Figure 6**) [34]. In the presence of chiral NHC generated from precursor **(5a*R*,10b*S*)-A1**, $RuCl_3$, and O_2, the oxidative reaction of cinnamaldehyde **2** with 2,4-pentanedione **31** were performed in 1,4-dioxane, affording lactone **32** in 98% yield with 93% ee. The proposed reaction mechanism involves the oxidation of Breslow intermediate, generated from NHC and enal **2**, by SET from $RuCl_3$. The second oxidation of radical cation intermediate by $RuCl_3$ gives α,β-unsaturated acyl azolium, which undergoes [3 + 3] annulation with 2,4-pentanedione **31**. In this catalysis, Ru(III) is regenerated through the oxidation of Ru(II) by molecular oxygen. Furthermore, cooperative catalysis using NHC and iridium catalyst was also developed [35, 36].

3. Cooperative NHC catalysis with photocatalysts

The combined use of NHC and photocatalyst has gained increasing attention as novel redox catalysis. The compatibility of NHC with ruthenium photocatalyst was demonstrated (**Figure 7**) [37]. Acylation of *N*-phenyltetrahydroisoquinoline **34** with butanal **33** was promoted by using chiral NHC, generated from precursor **(5a*S*,10b*R*)-A9** and photocatalyst $Ru(bpy)_3Cl_2$ in the presence of *m*-dinitrobenzene

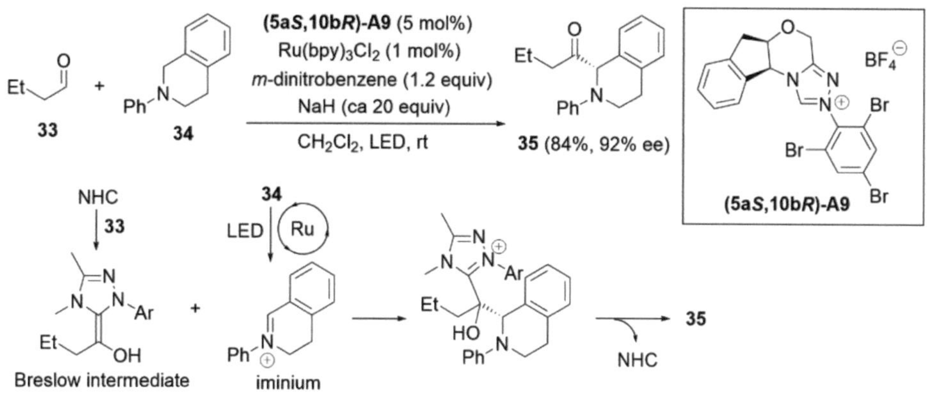

Figure 7.
Compatibility of NHC with ruthenium photoredox catalyst.

as an oxidant. This transformation proceeds *via* the generation of Breslow intermediate, which undergoes the addition to iminium intermediate generated by the photocatalytic oxidation of **34**. Finally, the release of free NHC catalyst results in the formation of acylated product **35**. The decarboxylative carbonylation reaction also proceeded *via* a similar mechanism [38].

The cooperative catalysis was applied to the oxidative transformation of aldehydes (**Figure 8**) [39–42]. The oxidative esterification of cinnamaldehyde **2** was achieved by the dual organocatalysis based on the cooperation between NHC and rhodamine 6G as an organophotocatalyst [39]. In this reaction, Breslow intermediate is photocatalytically oxidized to acyl azolium *via* the radical intermediate. The subsequent reaction of acyl azolium with MeOH gives ester **36**. Furthermore, the alkylation and esterification reaction of γ-oxidized enal **37** was developed [40]. When racemic precursor **A1** and photocatalyst Ru(bpy)$_3$(PF$_6$)$_2$ were employed, the reaction of γ-oxidized enal **37** with iodoacetonitrile **38** and MeOH gave γ-alkylated ester **39** in 86% yield. In this Ru-photocatalysis, iodoacetonitrile **38** acts as not only a radical source but also an oxidant. The oxidative Smiles rearrangement was also reported [42]. Under the cooperative catalysis conditions using NHC and 9-mesityl-10-methyl-acridin-10-ium as an organophotocatalyst, the oxidative Smiles rearrangement of *O*-aryl salicylaldehyde **40** proceeded effectively to give the aryl salicylate **41** in 79% yield. Initially, the photocatalytic oxidation of Breslow intermediate, generated from salicylaldehyde **40** to acyl azolium leads to the generation of the acid intermediate *via* the subsequent reaction of acyl azolium with H_2O. The subsequent oxidation of acid intermediate by photocatalysis promotes Smiles rearrangement to give another radical *via* the spirocyclic intermediate. Finally, the photocatalytic reduction of this radical gives the aryl salicylate **41**.

The cooperative catalysis for preparing ketones from carboxylic acid derivatives was studied (**Figure 9**) [43, 44]. The synthesis of ketone **44** was achieved by the combined NHC and Ir-photoredox catalysis of acyl imidazole **42** with benzyl Hantzsch ester **43** as a benzyl radical source [43]. In the presence of precursor **A12** (15 mol%), photocatalyst [Ir(dFCF$_3$ppy)$_2$(dtbpy)]PF$_6$ (1 mol%), and Cs_2CO_3, the reaction between imidazole **42** and Hantzsch ester **43** led to the formation of ketone **44** in 79% yield under the LED irradiation. In this catalysis, the iridium-photocatalyzed one-electron reduction of acyl azolium, generated from NHC and acyl imidazole **42**, affords a radical intermediate. This radical undergoes the subsequent radical-radical coupling with a benzyl radical generated by the iridium-photocatalyzed one-electron oxidation of benzyl Hantzsch ester **43**. The cooperative triple catalysis using NHC catalyst, Ru-photocatalyst, and sulfinate catalyst was

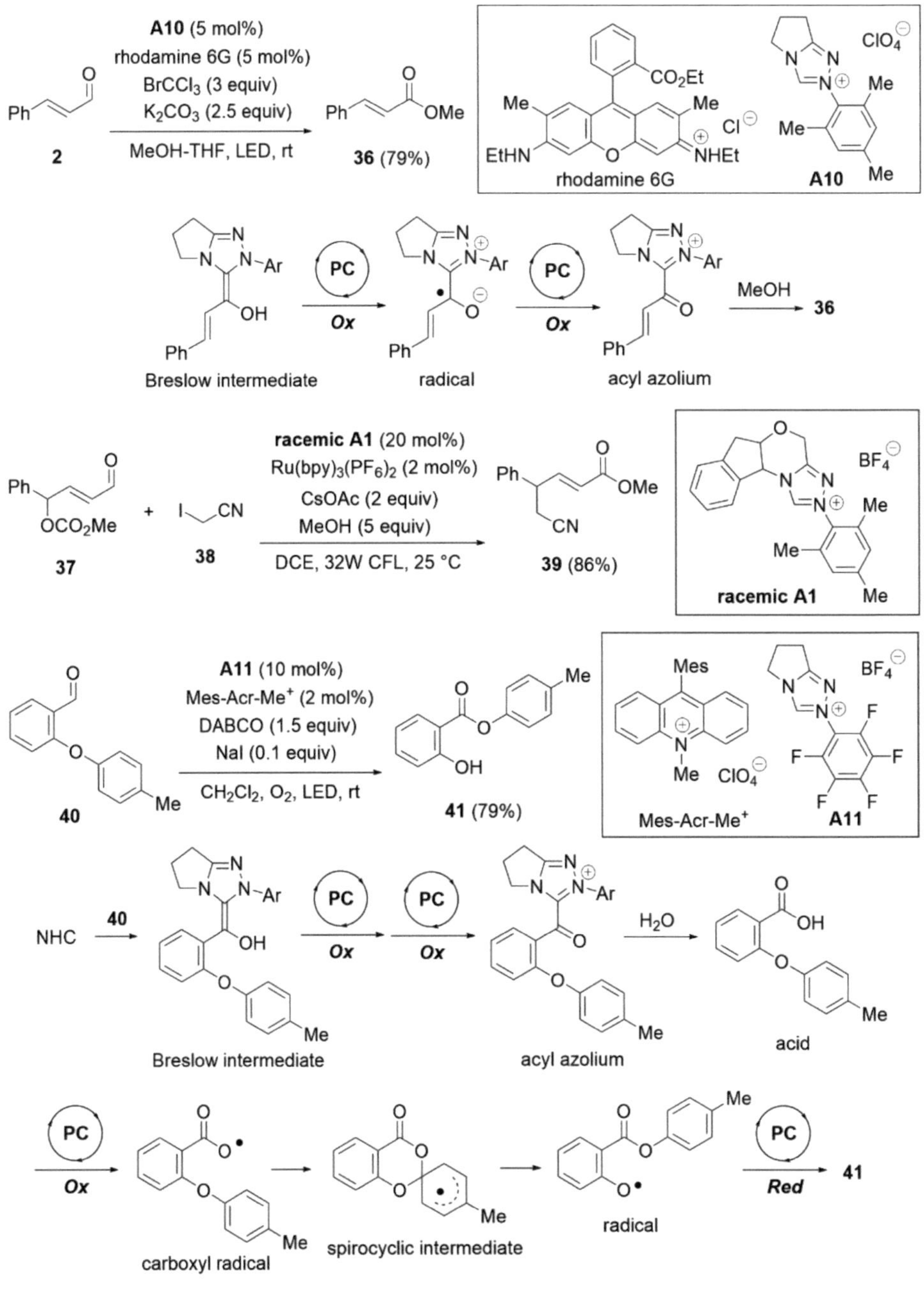

Figure 8.
Cooperative catalysis via oxidation of Breslow intermediates.

developed [44]. When precursor **A13** (15 mol%), Ru(bpy)$_3$(PF6)$_2$ (1.5 mol%), and 4-Cl-PhSO$_2$Na (25 mol%) were employed, the acylation of 4-methylstyrene **46** with benzoyl fluoride **45** was promoted under the CFL irradiation to give ketone **47** in 78% yield. This triple catalysis involves photocatalysis, NHC catalysis, and sulfinate catalysis. The NHC catalysis gives a ketyl radical *via* the photocatalytic reduction of acyl azolium, generated from benzoyl fluoride **45** and NHC. In the sulfinate catalysis, the photocatalytic oxidation of 4-Cl-PhSO$_2$Na affords sulfonyl radical, which adds to 4-methylstyrene **46** leading to the adduct radical. Finally, radical/radical cross-coupling between these two radicals leads to the acylated product **47**.

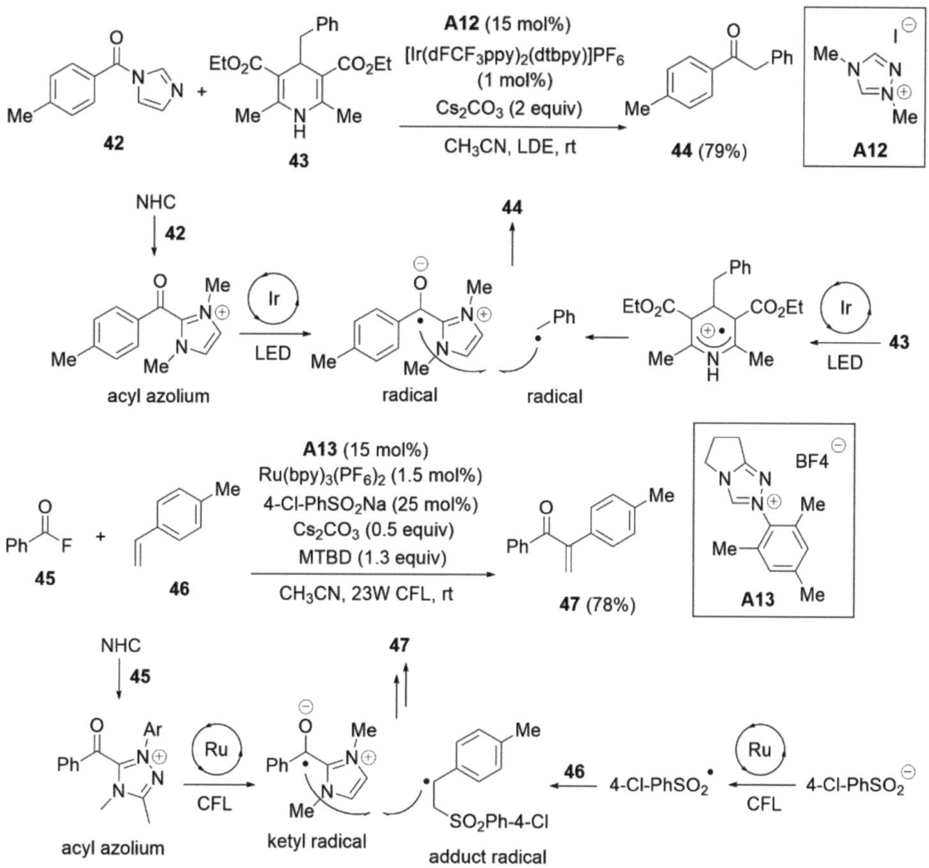

Figure 9.
Cooperative catalysis for preparing ketones.

In addition to the cooperative NHC catalysis with photocatalysts, the combined use of NHC catalysis and photoredox reaction in the absence of a photocatalyst has gained increasing attention as novel catalysis [45–47].

4. Conclusions

The N-heterocyclic carbenes (NHCs) are powerful and versatile organocatalysts that induce synthetically valuable chemical transformations. In addition to the cooperative catalysis using NHC/Lewis acid, NHC/Brønsted acid, and NHC/hydrogen-bonding organocatalyst, the cooperative NHC catalysis combined with transition-metal catalysts are emerging continuously. In the last few years, the combined use of NHC and photocatalyst has gained increasing attention as dual redox catalysis. The recent dramatic progress in NHC-induced cooperative catalysis disclosed a broader aspect of the utility of NHC-organocatalysis for synthetic organic chemistry. This chapter will inspire creative new contributions to organic chemists.

Acknowledgements

Our work was supported by JSPS KAKENHI Grant-in-Aid for Scientific Research (C) Grant Number 16K08188.

Author details

Hideto Miyabe
School of Pharmacy, Hyogo Medical University and Hyogo University of Health Sciences, Kobe, Japan

*Address all correspondence to: miyabe@huhs.ac.jp and miyabe@hyo-med.ac.jp

References

[1] Enders D, Niemeier O, Henseler A. Organocatalysis by *N*-heterocyclic carbenes. Chemical Reviews. 2007;**107**: 5606-5655. DOI: 10.1021/cr068372z

[2] Bugaut X, Glorius F. Organocatalytic umpolung: *N*-heterocyclic carbenes and beyond. Chemical Society Reviews. 2012;**41**:3511-3522. DOI: 10.1039/c2cs15333e

[3] Grossmann A, Enders D. *N*-Heterocyclic carbene catalyzed domino reactions. Angewandte Chemie International Edition. 2012;**51**:314-325. DOI: 10.1002/anie.201105415

[4] Vora HU, Wheeler P, Rovis T. Exploiting acyl and enol azolium intermediates *via* *N*-heterocyclic carbene-catalyzed reactions of α-reducible aldehydes. Advanced Synthesis & Catalysis. 2012;**354**:1617-1639. DOI: 10.1002/adsc.201200031

[5] Sarkar SD, Biswas A, Samanta RC, Studer A. Catalysis with *N*-heterocyclic carbenes under oxidative conditions. Chemistry A European Journal. 2013;**19**:4664-4678. DOI: 10.1002/chem.201203707

[6] Flanigan DM, Romanov-Michailidis F, White NA, Rovis T. Organocatalytic reactions enabled by *N*-heterocyclic carbenes. Chemical Reviews. 2015;**115**:9307-9387. DOI: 10.1021/acs.chemrev.5b00060

[7] Chen X-Y, Liu Q, Chauhan P, Enders D. *N*-Heterocyclic carbene catalysis *via* azolium dienolates: An efficient strategy for remote enantioselective functionalizations. Angewandte Chemie International Edition. 2018;**57**:3862-3873. DOI: 10.1002/anie.201709684

[8] Cohen DT, Scheidt KA. Cooperative Lewis acid/*N*-heterocyclic carbene catalysis. Chemical Science. 2012;**3**:53-57. DOI: 10.1039/c1sc00621e

[9] Wang MH, Scheidt KA. Cooperative catalysis and activation with *N*-heterocyclic carbenes. Angewandte Chemie International Edition. 2016;**55**:14912-14922. DOI: 10.1002/anie.201605319

[10] Liu Q, Chen X-Y. Dual *N*-heterocyclic carbene/photocatalysis: A new strategy for radical processes. Organic Chemistry Frontiers. 2020;**7**:2082-2087. DOI: 10.1039/d0qo00494d

[11] Nemoto T, Fukuda T, Hamada Y. Efficient synthesis of 3-substituted 2,3-dihydroquinolin-4-ones using a one-pot sequential multi-catalytic process: Pd-catalyzed allylic amination–thiazolium salt-catalyzed Stetter reaction cascade. Tetrahedron Letters. 2006;**47**:4365-4368. DOI: 10.1016/j.tetlet.2006.04.095

[12] Lebeuf R, Hirano K, Glorius F. Palladium-catalyzed *C*-allylation of benzoins and an NHC-catalyzed three component coupling derived thereof: Compatibility of NHC- and Pd-catalysts. Organic Letters. 2008;**10**:4243-4246. DOI: 10.1021/ol801644f

[13] He J, Tang S, Tang S, Liu J, Sun Y, Pan X, et al. Assembly of functionalized α-hydroxy carbonyl compounds *via* combination of *N*-heterocyclic carbene and Pd catalysts. Tetrahedron Letters. 2009;**50**:430-433. DOI: 10.1016/j.tetlet.2008.11.032

[14] Liu K, Hovey MT, Scheidt KA. A cooperative *N*-heterocyclic carbene/palladium catalysis system. Chemical Science. 2014;**5**:4026-4031. DOI: 10.1039/c4sc01536c

[15] Guo C, Fleige M, Janssen-Müller D, Daniliuc CG, Glorius F. Cooperative *N*-heterocyclic carbene/palladium-catalyzed enantioselective umpolung annulations. Journal of the American Chemical Society. 2016;**138**:7840-7843. DOI: 10.1021/jacs.6b04364

[16] Guo C, Janssen-Müller D, Fleige M, Lerchen A, Daniliuc CG, Glorius F. Mechanistic studies on a cooperative NHC organocatalysis/palladium catalysis system: Uncovering significant lessons for mixed chiral Pd(NHC)(PR_3) catalyst design. Journal of the American Chemical Society. 2017;**139**:4443-4451. DOI: 10.1021/jacs.7b00462

[17] Singha S, Patra T, Daniliuc CG, Glorius F. Highly enantioselective [5 + 2] annulations through cooperative *N*-heterocyclic carbene (NHC) organocatalysis and palladium catalysis. Journal of the American Chemical Society. 2018;**140**:3551

[18] Liu Y-J, Ding Y-L, Niu S-S, Ma J-T, Cheng Y. *N*-Heterocyclic carbene/palladium cascade catalysis: Construction of 2,2-disubstitiuted benzofuranones from the reaction of 3-(2-formylphenoxy)propenoates with allylic esters. Journal of Organic Chemistry. 2018;**83**:1913-1923. DOI: 10.1021/acs.joc.7b02849

[19] Ding Y-L, Zhao Y-L, Niu S-S, Wu P, Cheng Y. Asymmetric synthesis of multifunctionalized 2,3-benzodiazepines by a one-pot *N*-heterocyclic carbene/ chiral palladium sequential catalysis. Journal of Organic Chemistry. 2020;**85**:612-621. DOI: 10.1021/acs.joc.9b02693

[20] Gao J, Zhang J, Fang S, Feng J, Lu T, Du D. Synergistic *N*-heterocyclic carbene/palladium-catalyzed [3 + 2] annulation of vinyl enolates with 1-tosyl-2-vinylaziridine. Organic Letters. 2020;**22**:7725-7729. DOI: 10.1021/acs.orglett.0c02935

[21] Bai Y, Xiang S, Leow ML, Liu X-W. Dual-function Pd/NHC catalysis: Tandem allylation–isomerization–conjugate addition that allows access to pyrroles, thiophenes and furans. Chemical Communications. 2014;**50**:6168-6170. DOI: 10.1039/c4cc01750a

[22] Bai Y, Leng WL, Li Y, Liu X-W. A highly efficient dual catalysis approach for *C*-glycosylation: Addition of (*o*-azaaryl)carboxaldehyde to glycals. Chemical Communications. 2014;**50**: 13391-13393. DOI: 10.1039/c4cc06111j

[23] Bi W, Yang Y, Ye S, Wang C. Umpolung coupling of pyridine-2-carboxaldehydes and propargylic carbonates *via* *N*-heterocyclic carbene/ palladium synergetic catalysis. Chemical Communications. 2021;**57**:4452-4455. DOI: 10.1039/d1cc01311d

[24] Yasuda S, Ishii T, Takemoto S, Haruki H, Ohmiya H. Synergistic *N*-heterocyclic carbene/palladium-catalyzed reactions of aldehyde acyl anions with either diarylmethyl or allylic carbonates. Angewandte Chemie International Edition. 2018;**57**:2938-2942. DOI: 10.1002/anie.201712811

[25] Haruki H, Yasuda S, Nagao K, Ohmiya H. Dehydrative allylation between aldehydes and allylic alcohols through synergistic *N*-heterocyclic carbene/palladium catalysis. Chemistry A European Journal. 2019;**25**:724-727. DOI: 10.1002/chem.201805955

[26] Ohnishi N, Yasuda S, Nagao K, Ohmiya H. Synergistic *N*-heterocyclic carbene/palladium-catalyzed aldehyde acylation of allylic amines. Asian Journal of Organic Chemistry. 2019;**8**:1133-1135. DOI: 10.1002/ajoc.201900303

[27] Yang W, Ling B, Hu B, Yin H, Mao J, Walsh PJ. Synergistic *N*-heterocyclic carbene/palladium-catalyzed umpolung 1,4-addition of aryl iodides to enals. Angewandte Chemie International Edition. 2020;**59**:161-166. DOI: 10.1002/anie.201912584

[28] Ling B, Yang W, Wang Y-E, Mao J. Cooperative *N*-heterocyclic carbene/ palladium-catalyzed umpolung 1,4-addition of vinyl bromides to enals. Organic Letters. 2020;**22**:9603-9608. DOI: 10.1021/acs.orglett.0c03654

[29] Namitharan K, Zhu T, Cheng J, Zheng P, Li X, Yang S, et al. Metal and carbene organocatalytic relay activation of alkynes for stereoselective reactions. Nature Communications. 2014;**5**:3982. DOI: 10.1038/ncomms4982

[30] Zhang Z-J, Zhang L, Geng R-L, Song J, Chen X-H, Gong L-Z. *N*-Heterocyclic carbene/copper cooperative catalysis for the asymmetric synthesis of spirooxindoles. Angewandte Chemie International Edition. 2019;**58**:12190-12194. DOI: 10.1002/anie.201907188

[31] Zhou L, Wu X, Yang X, Mou C, Song R, Yu S, et al. Gold and carbene relay catalytic enantioselective cycloisomerization/cyclization reactions of ynamides and enals. Angewandte Chemie International Edition. 2020;**59**:1557-1561. DOI: 10.1002/anie.201910922

[32] Zhao J, Mück-Lichtenfeld C, Studer A. Cooperative *N*-heterocyclic carbene (NHC) and ruthenium redox catalysis: Oxidative esterification of aldehydes with air as the terminal oxidant. Advanced Synthesis & Catalysis. 2013;**355**:1098-1106. DOI: 10.1002/adsc.201300034

[33] Youn SW, Yoo HJ. One-pot sequential *N*-heterocyclic carbene/rhodium(III) catalysis: Synthesis of fused polycyclic isocoumarins. Advanced Synthesis & Catalysis. 2017;**359**:2176-2183. DOI: 10.1002/adsc.201700072

[34] Wang Q, Chen J, Huang Y. Aerobic oxidation/annulation cascades through synergistic catalysis of $RuCl_3$ and *N*-heterocyclic carbenes. Chemistry A European Journal. 2018;**24**:12806-12810. DOI: 10.1002/chem.201803254

[35] Zhang J, Gao Y-S, Gu B-M, Yang W-L, Tian B-X, Deng W-P. Cooperative *N*-heterocyclic carbene and iridium catalysis enables stereoselective and regiodivergent [3 + 2] and [3 + 3] annulation reactions. ACS Catalysis. 2021;**11**:3810-3821. DOI: 10.1021/acscatal.1c00081

[36] Singha S, Serrano E, Mondal S, Daniliuc CG, Glorius F. Diastereodivergent synthesis of enantioenriched α,β-disubstituted γ-butyrolactones *via* cooperative *N*-heterocyclic carbene and Ir catalysis. Nature Catalysis. 2020;**3**:48-54. DOI: 10.1038/s41929-019-0387-3

[37] DiRocco DA, Rovis T. Catalytic asymmetric α-acylation of tertiary amines mediated by a dual catalysis mode: *N*-heterocyclic carbene and photoredox catalysis. Journal of the American Chemical Society. 2012;**134**:8094-8097. DOI: 10.1021/ja3030164

[38] Du D, Zhang K, Ma R, Chen L, Gao J, Lu T, et al. Bio- and medicinally compatible α-amino-acid modification *via* merging photoredox and *N*-heterocyclic carbene catalysis. Organic Letters. 2020;**22**:6370-6375. DOI: 10.1021/acs.orglett.0c02202

[39] Yoshioka E, Inoue M, Nagoshi Y, Kobayashi A, Mizobuchi R, Kawashima A, et al. Oxidative functionalization of cinnamaldehyde derivatives: Control of chemoselectivity by organophotocatalysis and dual organocatalysis. Journal of Organic Chemistry. 2018;**83**:8962-8970. DOI: 10.1021/acs.joc.8b01099

[40] Dai L, Xia Z-H, Gao Y-Y, Gao Z-H, Ye S. Visible-light-driven *N*-heterocyclic carbene-catalyzed γ- and ε-alkylation with alky radicals. Angewandte Chemie International Edition. 2019;**58**:18124-18130. DOI: 10.1002/anie.201909017

[41] Dai L, Ye S. Photo/*N*-heterocyclic carbene co-catalyzed ring opening and γ-alkylation of cyclopropane enal. Organic Letters. 2020;**22**:986-990. DOI: 10.1021/acs.orglett.9b04533

[42] Xia Z-H, Dai L, Gao Z-H, Ye S. *N*-Heterocyclic carbene/photo-cocatalyzed oxidative smiles

rearrangement: Synthesis of aryl salicylates from *O*-aryl salicylaldehydes. Chemical Communications. 2020;**56**: 1525-1528. DOI: 10.1039/c9cc09272b

[43] Bay AV, Fitzpatrick KP, Betori RC, Scheidt KA. Combined photoredox and carbene catalysis for the synthesis of ketones from carboxylic acids. Angewandte Chemie International Edition. 2020;**59**:9143-9148. DOI: 10.1002/anie.202001824

[44] Liu K, Studer A. Direct α-acylation of alkenes *via N*-heterocyclic carbene, sulfinate, and photoredox cooperative triple catalysis. Journal of the American Chemical Society. 2021;**143**:4903-4909. DOI: 10.1021/jacs.1c01022

[45] Wang C, Wang Z, Yang J, Shi S-H, Hui X-P. Sequential visible-light and *N*-heterocyclic carbene catalysis: Stereoselective synthesis of tetrahydropyrano[2,3-*b*]indoles. Organic Letters. 2020;**22**:4440-4443. DOI: 10.1021/acs.orglett.0c01447

[46] Gao Z-H, Xia Z-H, Dai L, Ye S. *N*-Heterocyclic carbene catalyzed photooxidation: Intramolecular cross dehydrogenative coupling of tetrahydroisoquinoline-tethered aldehydes. Advanced Synthesis & Catalysis. 2020;**362**:1819-1824. DOI: 10.1002/adsc.202000164

[47] Mavroskoufis A, Rajes K, Golz P, Agrawal A, Ruß V, Götze JP, et al. *N*-Heterocyclic carbene-catalyzed photoenolization/Diels-Alder reaction of acid fluorides. Angewandte Chemie International Edition. 2020;**59**:3190-3194. DOI: 10.1002/anie.201914456

Chapter 3

N-Heterocyclic Carbenes (NHCs): An Introduction

Ruchi Bharti, Monika Verma, Ajay Thakur and Renu Sharma

Abstract

In 1991, the isolation and characterization of nitrogen heterocyclic carbene (NHCs) prompted the discovery of a new class of chemical compounds. NHCs have developed academic curiosity as one of the most potent tools in organic chemistry, exhibiting its utility in commercially relevant protocols. NHCs are cyclic compounds with a divalent carbon atom bonded to at least one nitrogen atom. The size of the carbene ring, the substituent moieties on the nitrogen atoms, and the extra atoms within the heterocycle can be changed to produce a variety of distinct NHCs with various electrical properties. They make excellent ligands in coordination chemistry because of their ability to act as donors and the consequent stable bonds with most transition metals. Free NHCs have also been used as organocatalysts in chemical reactions that require no metals. This chapter provides an outline of the N-Heterocyclic Carbenes in Contemporary Chemistry, including their general properties and highlighting the essential structural and electronic features of different NHCs along with their synthetic procedure.

Keywords: N-heterocyclic carbenes (NHCs), metal-organic framework, precursor, coordination chemistry, organocatalysis

1. Introduction

A carbene is a divalent neutral carbon-bearing six electrons in the valance shell & is considered very reactive to be isolated. Over the last 150 years, chemists have been fascinated about the carbenes and attempted for its isolation but failed [1]. Usually, the carbenes have a brief life span and play the role of very reactive intermediates. However, N-Heterocyclic carbene, in which the carbene center is settled on an N-heterocyclic ring, possesses different traits. It was first investigated in the early 1960s by Wanzlick [2, 3]. Shortly after that, in 1968, Wanzlick and Ofele reported the first application of NHC, where they function as a ligand to make complexes with metal [4, 5]. Later, the first crystalline NHC IAd (**Figure 1**) was first isolated and identified in 1991 by Arduengo et al., who encouraged a plethora of research on the transition metal complexes with NHC [6–19].

A broad range of NHCs is known with different carbene rings, substituting nitrogen atoms or additional heteroatom. The lifetime of NHCs is increased as the carbon is stabilized due to steric shielding. More generally, it can be said that this feature makes it a suitable fit ligand for coordination chemistry [20]. NHCs are also used as organocatalysts in metal-free chemical transformations [21]. This chapter discusses the basic outline of NHCs, including their standard structural features and properties, emphasizing various electronic and steric properties. In addition,

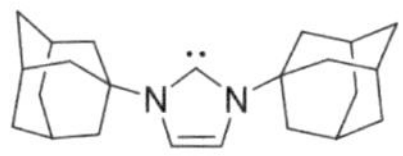
IAd (1c)

Figure 1.
First isolated NHC IAd (1c).

different synthetic routes leading to N-heterocyclic carbenes, along with their applications, have also conversed.

2. Chemistry of N-heterocyclic carbenes

2.1 Classes of NHCs and related stable carbenes

Many types of carbene compounds were labeled as NHCs in the past, and definitions of NHC based on their constitution are often subject to several interpretations [22]. N-heterocyclic carbenes can be defined as any chemical with a carbene center in a nitrogen-containing heterocyclic ring. A broad array of carbenes depending on substituents, size of the ring, and degree of stabilization of heteroatom are available, out of which a few important selected ones are shown in **Figure 2**. Another important classes of NHCs are imidazolinylidene (1), Tetrahydropyrimidinylidene (2), *N,N*-Diamidocarbene (DAC) (3), Benzimidazolylidene (4). These NHCs trigger small molecules like NH_3 & can go for insertion into alkenes [23]. Out of these Imidazolylidene-derived NHCs (1) like IPr or IDipp (1a), IMes (1b), and IAd (1c) is the commonest NHCs that are mainly used as ligands with block elements. They function as catalysts in cross-coupling reactions or other significant organic–inorganic transformations [24, 25]. One more category of NHCs was reported by Bertrand et al. [26] where he reported only one nitrogen-containing NHCs like pyrrolidinylidene usually termed as cyclic on the basis of nature of the substituent present neighboring to the carbene center [27, 28].

These compounds are more attracted to the π-electron cloud than other categories of NHCs and help stabilize delicate representative elements and organic free radicals [29]. Another category of NHCs with more than two nitrogen atoms in their heterocyclic framework is also present. Triazolylidenes (8) are examples of this class, widely used as organocatalysts for many transformations [30]. NHC classes having a different heteroatom like oxygen [oxazolylidene (6)], sulfur thiazolylidene (7) are also accessible. Different NHCs moieties can be obtained by developing the carbine center at different positions. In usual NHCs, the carbene center is generally present between the two nitrogen atoms as in imidazolylidenes **1**, generating the carbene center at the 4-position. However, that species is labeled mesoionic or abnormal carbene (MIC or an NHC), which is not neutral, and it is not possible to draw its non-zwitterionic resonance structure (9) [31, 32]. It is termed as "remote" NHC (rNHC) (10) when the carbene center is away from the nitrogen atom [33].

All aforementioned act as NHCs, but some related non-NHCs have also been reported, exhibiting the same characteristics as shown in **Figure 3** [34]. One of these is acyclic diamino carbenes (ADCs) (11) [35–37]. A carbene species 12 stabilized by adjoining silicon & phosphorus substituent was reported by Bertrand in 1988 and after 3 years free NHC was also isolated [38]. Cyclopropenylidene compounds (13) with exocyclic nitrogens were also synthesized by Bertrand [39] while "bent allene" species (14) have themselves been used as ligands [40]. These compounds are considered as NHCs stabilized acyclic carbene.

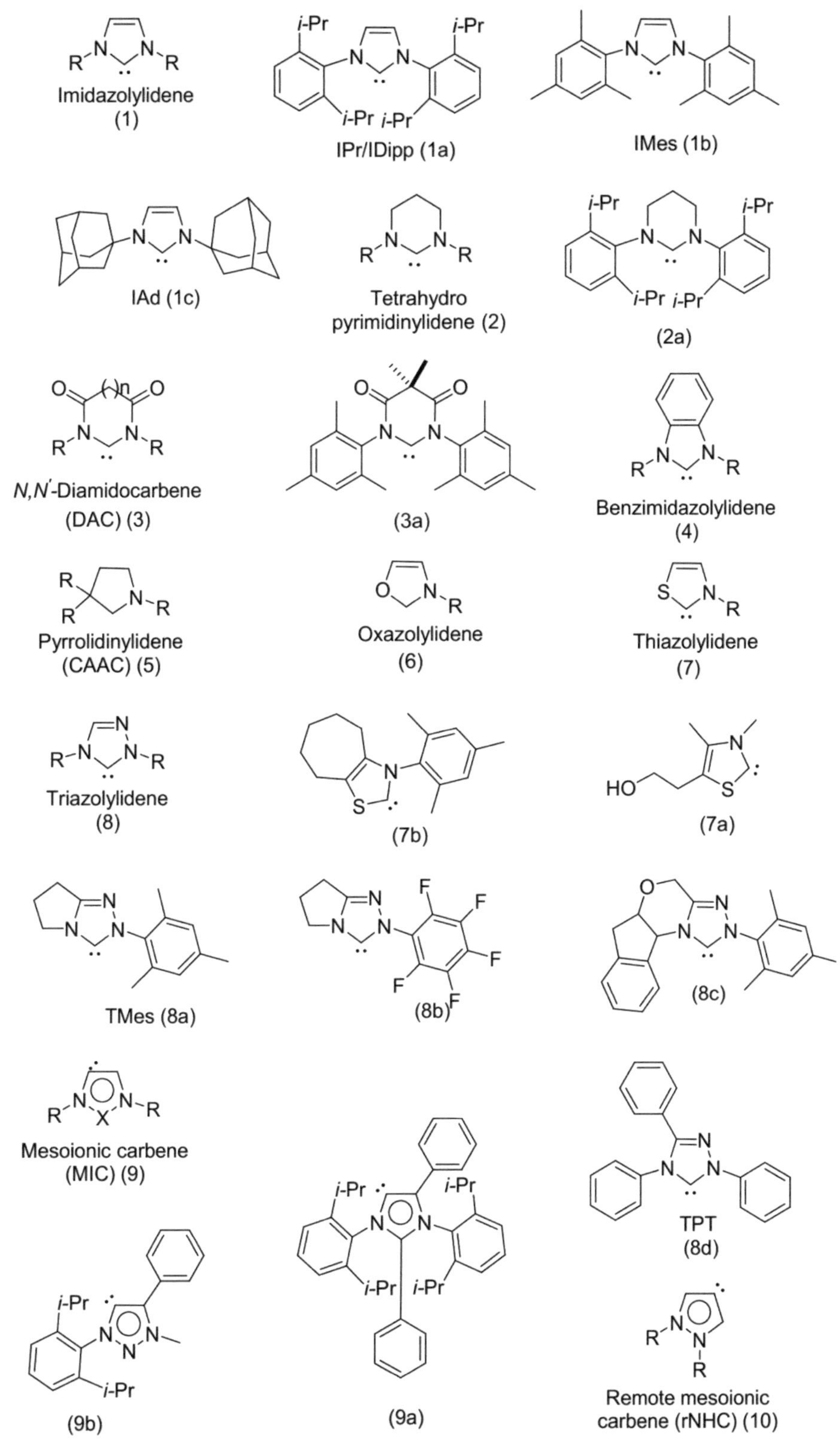

Figure 2.
Some examples of important classes of NHC.

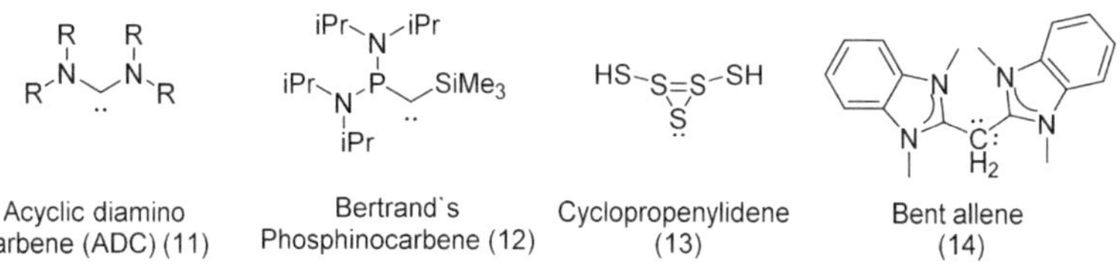

Figure 3.
Related classes of stable carbene.

2.2 Common structural features of NHCs

Nitrogen heterocycles are fundamental in defining the characteristics of NHC. Ranging from four-membered heterocycles, NHCs can be poly NHCs also, but the five-membered NHCs are most common (**Figure 4**), and all are based on imidazole and imidazolydiene [41]. Several structural features of imidazolylidene (1) are similar in all variants of NHCs, which helps stabilize the carbenes. Alteration in the structure and substituents of imidazolylidene develops diverse behavior in NHCs.

In the basic structure of imidazolylidene, the carbene carbon is attached to the two nitrogen atoms of the heterocycle. The aliphatic/aromatic substituent on the nitrogen atom (s) is denoted as the N-substituent(s). The remaining positions, i.e., the 4- and 5-positions in imidazolylidenes, are referred to as the NHC backbone. The substituents present on the ring backbone do not contribute any steric effect at the carbene center and affect the electronic environment only. However, substituents present on the nitrogen of the ring greatly influence the steric properties of the carbene center (**Figure 5**).

2.2.1 Stabilization of the carbene center

The steadiness of the carbene center in NHCs can be attributed to kinetic and thermodynamic factors. A carbene is carbon with an incomplete octet formed as an intermediate but does not undergo dimerization to form an alkene. Similarly, NHC bearing bulky aryl or alkyl substituent on nitrogen atom does not dimerize due to steric clashing and is termed the Wanzlick equilibrium. Electronic properties of NHC further contribute toward stability as they bear a singlet ground state despite classical carbene, which carries a triplet ground state. The lone pair of electrons of NHCs (singlet) is confined in sp^2-hybridized orbital, which exists in the plane of the ring in the highest occupied molecular orbital and an empty p-orbital is lying

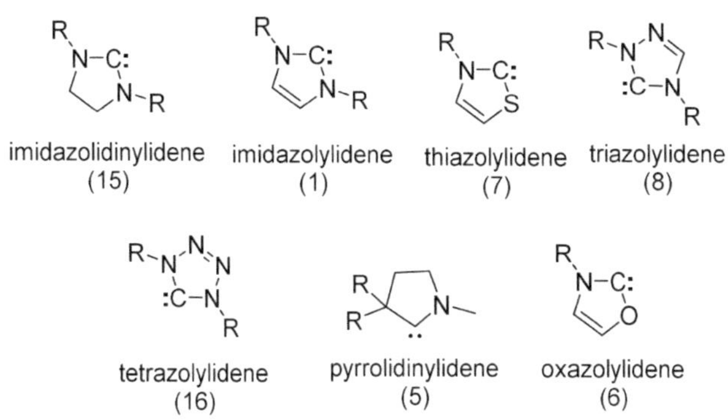

Figure 4.
Common five-membered heterocyclic rings.

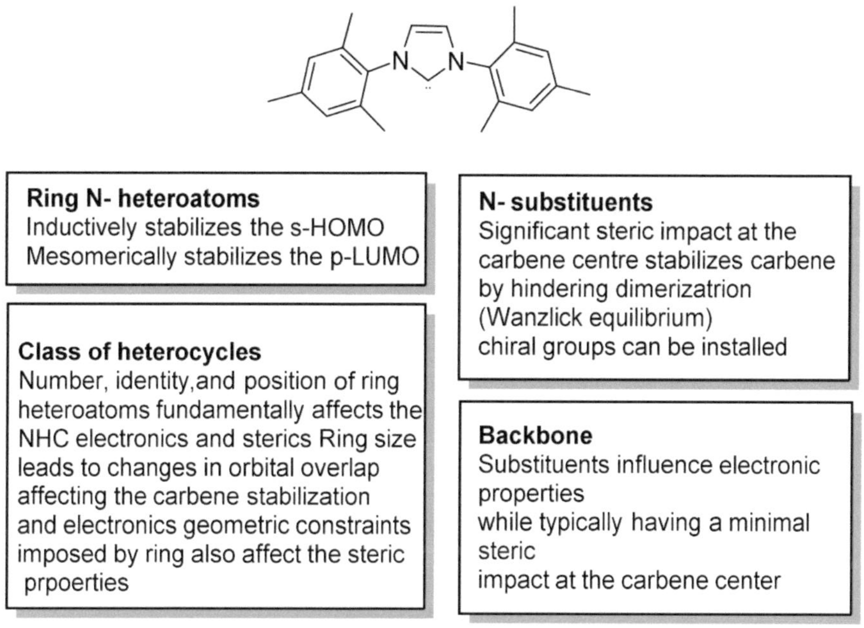

Figure 5.
General structural features of NHCs(1c).

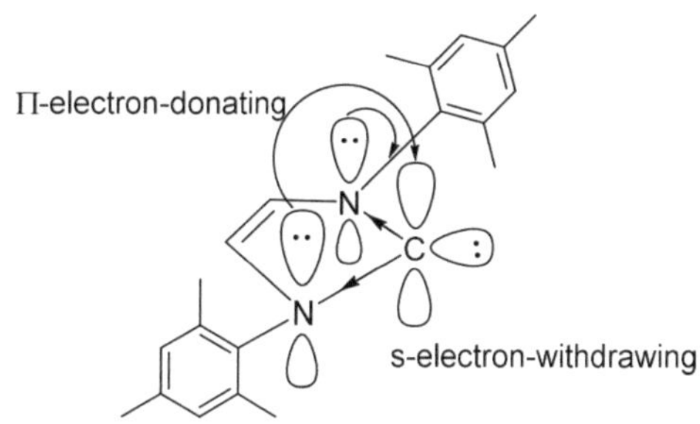

Figure 6.
Stabilization of the carbene by adjacent ring nitrogens.

perpendicular to it in the lowest unoccupied molecular orbital (**Figure 6**). As reported by Goddard et al. [42] singlet carbenes do not undergo dimerization easily, and adjacent nitrogen in NHCs further decreases the energy of HOMO, which causes sizeable singlet-triplet energy gaps. Interrelation among the lone pair of electrons of nitrogen present in p-orbital with empty LUMO (Mesomeric donation) also contributes to stabilizing the structure of the singlet ground-state.

2.3 Attractive features of N-heterocyclic carbenes

NHCs have gained much application as ligands for transition metal catalysis due to their steric and electronic behavior leading toward complex stability.

2.3.1 Electronic character

N-Heterocyclic carbenes belong to the category of very electron-rich ligands, although their degree of π-acceptor power is still doubtful. The electron donation property of NHC depends on the nature/type of metal, the substituents, and the

Ligand	^{v}CO (A_1) (cm^{-1})	^{v}CO (E) (cm^{-1})
IMes	2050.7	1969.8
SIMes	2051.5	1970.6
IPr	2051.5	1970.0
SIPr	2052.2	1971.3
Icy	2049.6	1964.6
PCy_3	2056.4	1973
PPh_3	2068.9	1990
$PtBu_3$	2056.1	1971

Table 1.
IR values for the carbonyl stretching frequencies in $LNi(CO)_3$ measured in CH_2Cl_2.

co-ligands present on the NHC ring corresponding to the metal [43–46]. The capability to donate electrons can be calculated by comparing the stretching frequencies of CO ligands of complexes like $LRh(CO)_2Cl$, [47], $LIr(CO)_2Cl$ [47] with L = NHC or $LNi(CO)_3$ [48]. Hence, it is evident that N-heterocyclic carbenes are electron-rich ligands than the most basic trialkyl phosphines (**Table 1**) [49].

Furthermore, the NHCs have very similar levels of electron-donating ability as compared to the phosphines. The reason for this difference can be explained as the substituents of NHCs are exchanged only on the periphery of the ligand while for phosphines the different substituents are directly attached to the donor atom itself. Therefore, the finest way is to modify the electronic behavior of an NHC is to alter the type of the azole ring. In this way, it is rational to believe that the order of the electron-donating capability is benzimidazole < imidazole < imidazoline. It is easy to understand that components with +I and +M-effects increase the electron density of a NHC making it a better donor, while components with −I and −M-effects show the opposite behavior. However, some substituents have opposing electronic effects and complicate the situation. For example, halo groups (F−, Cl−, Br−, and I−) bound to the carbon atom exhibit the –I-effect due to their increased electronegativity but they also have the +M-effect as a result of three loan pairs for donation that need to be considered as well [50].

This electron-rich property of NHCs impacts many rudimentary levels of the catalytic process, e.g. smoothing the oxidative addition step. Hence, the complex of NHC with metal are suitable for cross-coupling reactions of non-activated aryl chlorides, which encounter the catalyst with a challenging oxidative addition step [51].

2.3.2 Sterics

NHCs are often used as phosphine mimics, but both structures are pretty different (**Figure 7**). Phosphine complexes have a cone-like structure where alkyl/aryl groups are pointed away from phosphorous. So, the steric properties of NHCs can be elucidated using Tolman's ingenious cone angle descriptors [49].

Figure 7.
Shape of phosphines and NHC.

The topology of N-heterocyclic carbene is contrary to phosphene and is more complex to predict factors determining its steric effect. The shape of NHC is defined by the position of the alkyl/aryl group present on the nitrogen(s) of the heterocycle. NHCs have been featured as fence- or fan-like [52]. The side groups are bent toward the metal and wrap it by forming a pocket (**Figure 7**). The steric and electronic properties of NHCs can change via rotation around the metal-carbene bond, hence making it anisotropic.

2.3.3 NHCs as ligands

Most of the metals form a very stable bond with Nitrogen heterocyclic carbene [46, 53]. Whereas quite same bond dissociation energies have been noticed for unsaturated & saturated NHCs with the same steric impacts, phosphines generally form weaker bonds (**Table 2**) [28, 54].

Consequently, the equilibrium between the carbene metal complex and free carbene exists toward the complex compared to phosphines (**Figure 8**). It increases the lifetime of complex but still N-heterocyclic carbenes very sensitive & reactive for many electron-loving moieties. They need careful isolation and storage. The consequential unusual firmness of NHC-metal complexes has been explored in many demanding protocols such as coupling reactions [55, 56], polymerization [57, 58], transfer hydrogenation [59–61], photocatalysis [62, 63], and many other [64–70].

However, escalating publications reveal that the bond between metal and carbene is not unreactive [53, 71–74]. As seen during the migratory insertion of an NHC into the double bond of ruthenium-carbon [75] removal of alkyl imidazolium salts from NHC alkyl complexes via reductive elimination, [76] or the ligand substitution of NHC ligands by phosphines, [77, 78]. Additionally, during applications

Ligand	$\%V_{Bur}$ for M-L (2.00 Å)	BDE[kcal/mol] (theoretical) for L in $Ni(CO)_3L$
IMes	26	41.1
SIMes	27	40.2
IAd	37	20.4
I*t*Bu	37	24.0
PPh_3	27	26.7

Table 2.
Steric demand and bond strength of some important ligands.

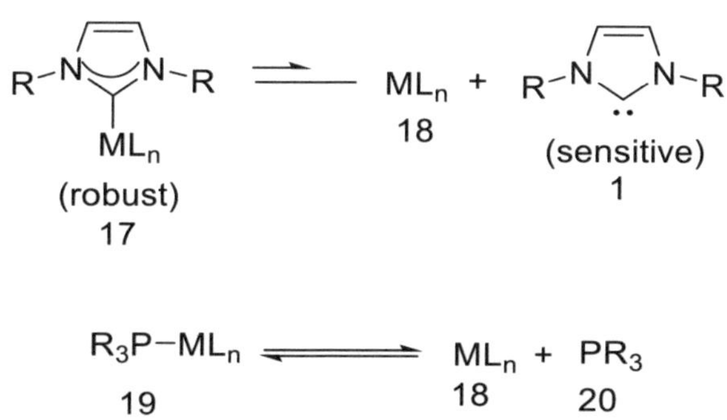

Figure 8.
Equilibrium of complexation.

of palladium NHC complexes, the generation of palladium black is observed, which points toward the decomposition pathways.

2.3.4 NHC adducts-bonding

The researchers have investigated the bonding between carbene center and metals/non-metals over the past two decades [8, 15]. The main feature of this bonding is the donation of carbene lone pair of electrons into empty sigma orbital of metal/non-metal. It is evident in the graphical representations (**Figure 9**) of NHC-metal complexes (53, 54, and 55) that inspite of a double bond usually a single bond is shown. Therefore, in the deficiency of steric restrictions or chelation, the bond between metal and NHC can rotate easily. The π-orbital in the NHC's complexes and metal is restricted within the NHC ring so indicated by a curved line between the heteroatoms of the ring.

Metal donates its electrons in the carbene's vacant p-orbital (LUMO) of carbene. The carbene can also donate electrons in the vacant π-orbitals of metal. The significance of each factor depends on the nature and geometry of another ligand on the complex and, most importantly nature of carbene itself. Using spectroscopic methods π-accepting ability of different NHCs can be measured and quantified [54, 79].

3. Synthesis of NHCs

As a broad range of N-heterocyclic carbene is based on 5-membered rings, hence the simplest way to prepare them is via removal of a proton from the related azolium salts, such as imidazolium, pyrazolium, triazolium, tetrazolium, benzimidazolium, oxazolium, or thiazolium salts by using appropriate bases. The

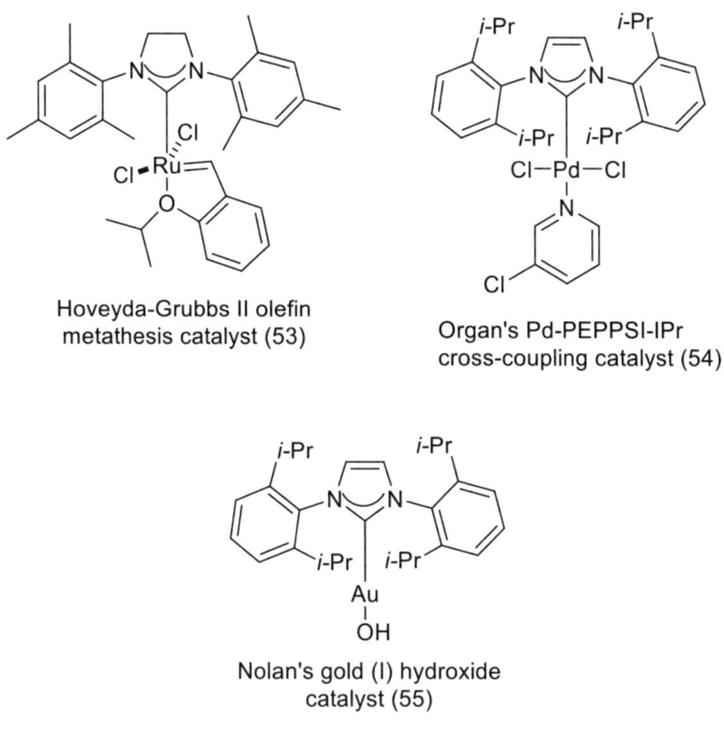

Hoveyda-Grubbs II olefin metathesis catalyst (53)

Organ's Pd-PEPPSI-IPr cross-coupling catalyst (54)

Nolan's gold (I) hydroxide catalyst (55)

Figure 9.
NHC-containing transition metal catalysts.

value of dissociation constant (pKa) of benzimidazolium and imidazolium salts was observed between 21 and 24, giving them a place in between the neutral carboxylic acid, acetone, and ethylacetate [9, 15, 80, 81]. Imidazolium salts are synthesized by following two routes. First method involves alkylation of existing imidazoles using appropriate electrophiles resulting in the generation of N-alkyl-substituted imidazolium salts. Whereas in other methods the imidazolium ring is synthesized by condensation reactions (**Figures 10** and **11**). As the attention for the synthesis of

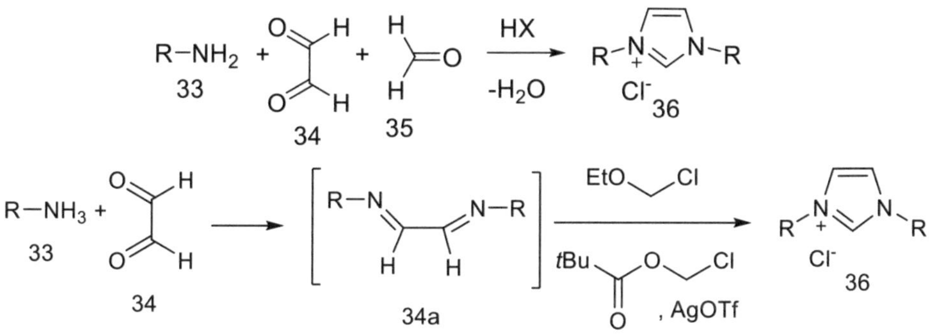

Figure 10.
Symmetric NHCs Synthesis.

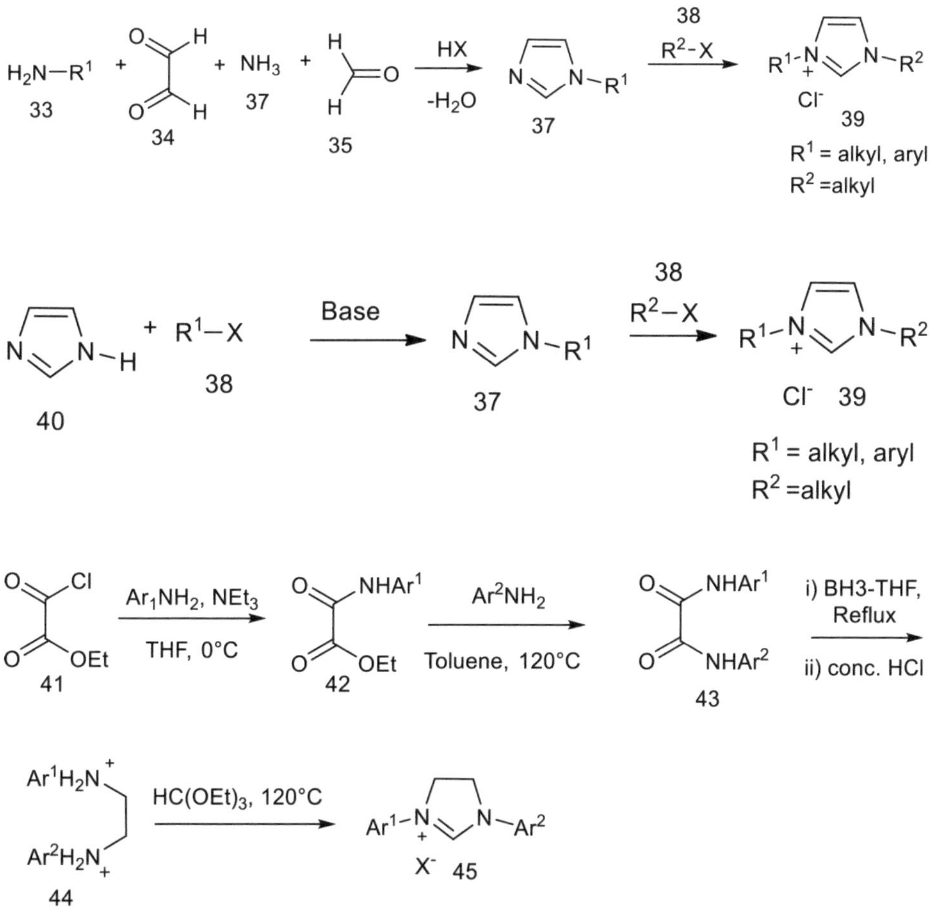

Figure 11.
Unsymmetrical Synthesis of NHCs.

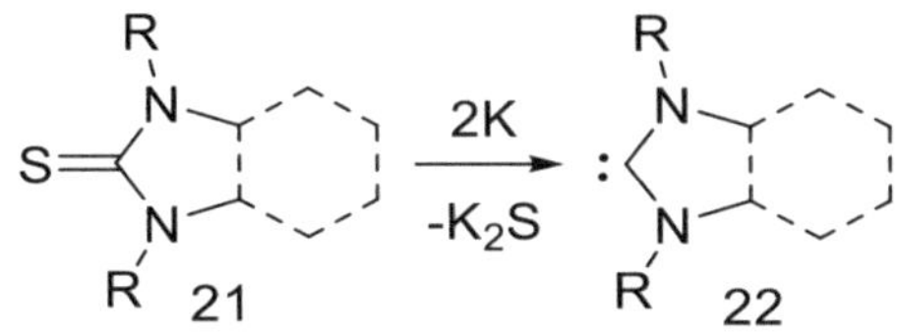

Figure 12.
Desulfurization of thiones.

NHCs and imidazolium salts is increasing very rapidly, the methodologies related to its synthesis have been improved regularly.

a. **Reductive desulfurization of N-heterocyclic thiones**: a variety of-heterocyclic carbene such as saturated, unsaturated, and benzannulated can be prepared by using this method. Imidazole-2-thione gets reduced to carbene by using potassium in boiling THF for 4 h (**Figure 12**). At the same time, benzimidazole-2-thiones undergo a reduction in the presence of Na/K alloy along with toluene within three weeks [82–86].

b. **α-Elimination or dehalogenation:** Imidazolinium salts derivatives bearing bulky *N*, *N′*-substituents **23** undergo α-elimination reaction to form the NHCs **24** under the influence of thermal induction (**Figure 13a**) [30].

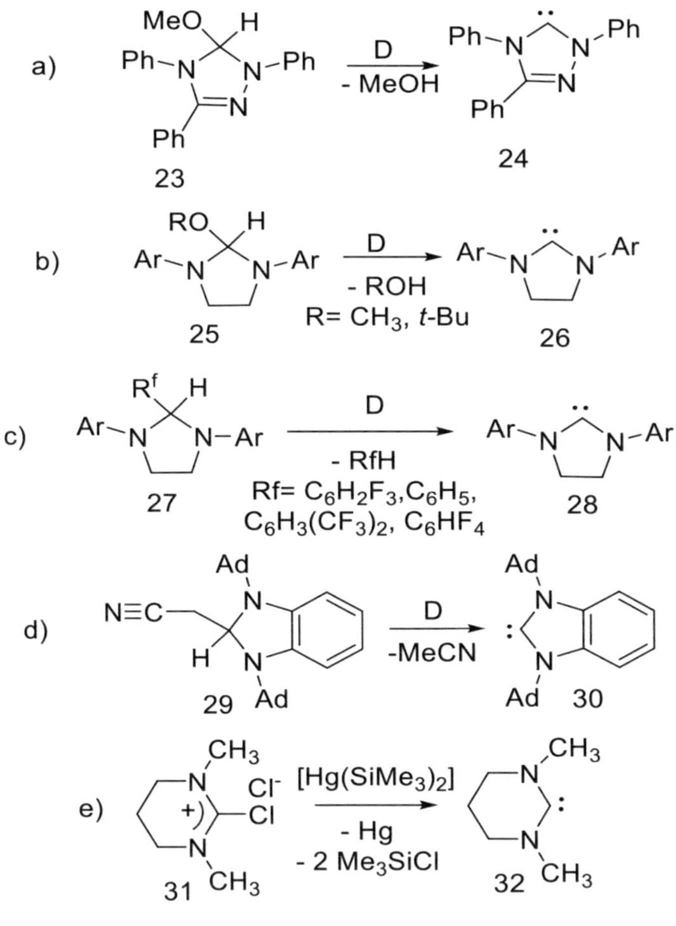

Figure 13.
NHCs synthesis (a-e) by α-Elimination or dehalogenation.

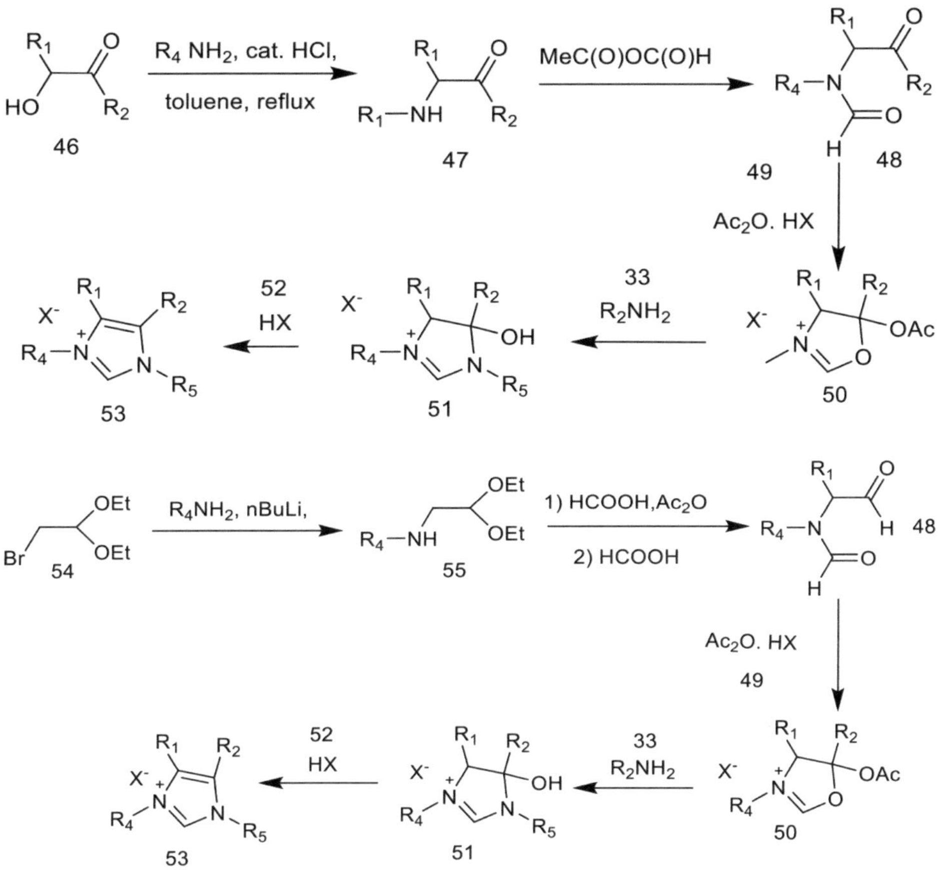

Figure 14.
Unsymmetrical synthesis of imidazolium salts.

The corresponding alcohol elimination from 2-alkoxyimidazolidines **25** to give imidazolin-2-ylidenes of type **26** (**Figure 13b**) was outlined by Grubbs [87] after early unsuccessful attempts by Wanzlick and Kleiner [88]. Imidazoline-2-ylidenes **26** were also accessible via α-elimination of fluorinated aryls from 2-(fluorophenyl)imidazolines **27** (**Figure 13c**) [89, 90]. The α-elimination of acetonitrile from **29** to yield the benzimidazole-2-ylidene **30** has also been described (**Figure 13d**) [91]. Bertrand reported the dechlorination reaction among tetrahydropyrimidiniumchloride **31** and bis (trimethylsilyl)mercury leading to NHC **32** (**Figure 13e**) [92].

c. **Symmetric synthesis**: For symmetric NHCs (36), synthesis formaldehyde (35) is treated with primary amine (33) and glyoxal (34) under strongly acidic conditions. Otherwise, the bisimine intermediate (34a) is reacted with electrophilic C1-fragments (chloromethyl ethyl ether or chloromethylpivalate) after isolation [81, 93–95]. During few crucial methodologies, the addition of stoichiometric amounts of AgOTfwas evidenced to be helpful (**Figure 10**) [96].

d. **Unsymmetrical synthesis:** Unsymmetrically, NHCs are prepared by adding alkyls to monosubstitutedimidazoles (**Figure 11**) [96–103]. The unsymmetrical imidazolium salt (39), so obtained is incorporating a counter anion (X^-) to maintain the electric neutrality but careful selection of the counter anion is sensible as it affects the dissolution of the imidazolium salt. It is noteworthy that non-coordinating counter ions like OTf- or BF_4-increase the solubility of the salts.

In another protocol for synthesizing unsymmetrical NHCs, the α-hydroxy ketone (46) is reacted with the amine (R_4-NH_2) of choice under acidic conditions. Thus, by azeotropic removal of water, the α-aminoketones (47) are obtained, which are then N-formylated (48) before reaction with $(CH_3CO)_2O$ in the presence of a aq. HBF_4 or $HClO_4$. The subsequent cyclization stops at the intermediate acetal stage. Also, the reaction of the oxazolinium adducts 49 with R_2–NH_2 goes on easily, generating hydroxylateimidazolinium salts bromoacetaldehyde diethyl acetal (53) (**Figure 14**) [104].

4. Conclusions

Since the 1960s, when the initial studies on NHC began, and later (1991) when Arduengo isolated the first NHC, the applications of these compounds are growing across various domains of chemistry. Nature itself has chosen an NHC as a part of vitamin B1 to achieve organocatalytic reactions in vivo. Over the past few years, NHCs have unfolded novel categories of enantioselective organic transformations. This chapter provides a brief outline of the fundamental properties and synthesis of NHCs by analyzing different methods used to quantify their steric & electronic properties.

Acknowledgements

The authors are thankful to Chandigarh University for providing generous support for this work.

Conflict of interest

The authors declare no conflict of interest.

Author details

Ruchi Bharti*, Monika Verma, Ajay Thakur and Renu Sharma
Department of Chemistry, University Institute of Sciences, Chandigarh University, Mohali, Punjab, India

*Address all correspondence to: ruchi.uis@cumail.in

References

[1] Bertrand G. Carbene Chemistry: From Fleeting Intermediates to Powerful Reagents. Boca Raton, Florida, United States: CRC Press; 2002

[2] Benhamou L, Chardon E, Lavigne G, Bellemin-Laponnaz S, Cesar V. Synthetic routes to N-heterocyclic carbene precursors. Chemical Reviews. 2011;**111**(4):2705-2733

[3] Wanzlick HW. Aspects of nucleophilic carbene chemistry. Angewandte Chemie International Edition in English. 1962;**1**(2):75-80

[4] Wanzlick HW, Schönherr HJ. Direct synthesis of a mercury salt-carbene complex. Angewandte Chemie International Edition in English. 1968; 7(2):141-142

[5] Öfele K. 1, 3-Dimethyl-4-imidazolinyliden-(2)-pentacarbonyl chromeinneuer Übergangsmetall-carben-komplex. Journal of Organometallic Chemistry. 1968;**12**(3):42-43

[6] Arduengo AJ III, Harlow RL, Kline M. A stable crystalline carbene. Journal of the American Chemical Society. 1991;**113**(1):361-363

[7] Crudden CM, Allen DP. Stability and reactivity of N-heterocyclic carbene complexes. Coordination Chemistry Reviews. 2004;**248**(21–24):2247-2273

[8] Díez-González S, Nolan SP. Stereoelectronic parameters associated with N-heterocyclic carbene (NHC) ligands: A quest for understanding. Coordination Chemistry Reviews. 2007; **251**(5–6):874-883

[9] Colacino E, Martinez J, Lamaty F. Preparation of NHC–ruthenium complexes and their catalytic activity in the metathesis reaction. Coordination Chemistry Reviews. 2007;**251**(5–6): 726-764

[10] Lee HM, Lee CC, Cheng PY. Recent development of functionalized N-heterocyclic carbene ligands: Coordination chemistry and catalytic applications. Current Organic Chemistry. 2007;**11**(17):1491-1524

[11] Kühl O. The chemistry of functionalized N-heterocyclic carbenes. Chemical Society Reviews. 2007;**36**(4): 592-607

[12] Lin IJ, Vasam CS. Preparation and application of N-heterocyclic carbene complexes of Ag (I). Coordination Chemistry Reviews. 2007;**251**(5–6): 642-670

[13] Cavell K. N-heterocyclic carbenes/ imidazolium salts as substrates in catalysis: The catalytic 2-substitution and annulation of heterocyclic compounds. Dalton Transactions. 2008; **47**:6676-6685

[14] Marion N, Nolan SP. Well-defined N-heterocyclic carbenes– palladium (II) precatalysts for cross-coupling reactions. Accounts of Chemical Research. 2008;**41**(11):1440-1449

[15] Jacobsen H, Correa A, Poater A, Costabile C, Cavallo L. Understanding the M (NHC)(NHC= N-heterocyclic carbene) bond. Coordination Chemistry Reviews. 2009;**253**(5–6):687-703

[16] Wang F, Liu LJ, Wang W, Li S, Shi M. Chiral NHC–metal-based asymmetric catalysis. Coordination Chemistry Reviews. 2012;**256**(9–10): 804-853

[17] Liu W, Gust R. Metal N-heterocyclic carbene complexes as potential antitumor metallodrugs. Chemical Society Reviews. 2013;**42**(2):755-773

[18] Nelson DJ, Nolan SP. Quantifying and understanding the electronic properties of N-heterocyclic carbenes.

Chemical Society Reviews. 2013;**42**(16): 6723-6753

[19] Hu C, Li X, Wang W, Zhang R, Deng L. Metal-N-heterocyclic carbene complexes as antitumor agents. Current Medicinal Chemistry. 2014;**21**(10): 1220-1230

[20] Diez-Gonzalez S, Marion N, Nolan SP. N-heterocyclic carbenes in late transition metal catalysis. Chemical Reviews. 2009;**109**(8):3612-3676

[21] Enders D, Niemeier O, Henseler A. Organocatalysis by N-heterocyclic carbenes. Chemical Reviews. 2007; **107**(12):5606-5655

[22] Hopkinson MN, Glorius F. An overview of NHCs. N-Heterocyclic Carbenes in Organocatalysis. 1st ed. 2018. pp. 1-35

[23] Moerdyk JP, Schilter D, Bielawski CW. N, N′-diamidocarbenes: Isolable divalent carbons with bona fide carbene reactivity. Accounts of Chemical Research. 2016;**49**(8): 1458-1468

[24] Shi S, Nolan SP, Szostak M. Well-defined palladium (II)–NHC precatalysts for cross-coupling reactions of amides and esters by selective N–C/O–C cleavage. Accounts of Chemical Research. 2018;**51**(10):2589-2599

[25] Ohmiya H. N-heterocyclic carbene-based catalysis enabling cross-coupling reactions. ACS Catalysis. 2020;**10**(12): 6862-6869

[26] Lavallo V, Canac Y, Präsang C, Donnadieu B, Bertrand G. Stable cyclic (alkyl)(amino) carbenes as rigid or flexible, bulky, electron-rich ligands for transition metal catalysts: A quaternary carbon atom makes the difference. Angewandte Chemie International Edition. 2005;**44**(35):5705-5709

[27] Soleilhavoup M, Bertrand G. Cyclic (alkyl)(amino) carbenes (CAACs): Stable carbenes on the rise. Accounts of Chemical Research. 2015;**48**(2):256-266

[28] Rao B, Tang H, Zeng X, Liu LL, Melaimi M, Bertrand G. Cyclic (amino)(aryl) carbenes (CAArCs) as strong σ-donating and π-accepting ligands for transition metals. Angewandte Chemie International Edition. 2015;**54**(49): 14915-14919

[29] Martin CD, Soleilhavoup M, Bertrand G. Carbene-stabilized main group radicals and radical ions. Chemical Science. 2013;**4**(8):3020-3030

[30] Enders D, Breuer K, Raabe G, Runsink J, Teles JH, Melder JP, et al. Preparation, structure, and reactivity of 1, 3, 4-Triphenyl-4, 5-dihydro-1H-1, 2, 4-triazol-5-ylidene, a new stable carbene. Angewandte Chemie International Edition in English. 1995; **34**(9):1021-1023

[31] Aldeco-Perez E, Rosenthal AJ, Donnadieu B, Parameswaran P, Frenking G, Bertrand G. Isolation of a C5-deprotonated imidazolium, a crystalline "abnormal" N-heterocyclic carbene. Science. 2009;**326**(5952): 556-559

[32] Ghadwal RS. Carbon-based two-electron σ-donor ligands beyond classical N-heterocyclic carbenes. Dalton Transactions. 2016;**45**(41): 16081-16095

[33] Schuster O, Yang L, Raubenheimer HG, Albrecht M. Beyond conventional N-heterocyclic carbenes: Abnormal, remote, and other classes of NHC ligands with reduced heteroatom stabilization. Chemical Reviews. 2009; **109**(8):3445-3478

[34] Melaimi M, Soleilhavoup M, Bertrand G. Stable cyclic carbenes and related species beyond diaminocarbenes. Angewandte Chemie International Edition. 2010;**49**(47): 8810-8849

[35] Vignolle J, Cattoen X, Bourissou D. Stable noncyclic singlet carbenes. Chemical Reviews. 2009;**109**(8): 3333-3384

[36] Igau A, Grutzmacher H, Baceiredo A, Bertrand G. Analogous. alpha.,. alpha.'-bis-carbenoid, triply bonded species: synthesis of a stable. lambda. 3-phosphino carbene-. lambda. 5-phosphaacetylene. Journal of the American Chemical Society. 1988; **110**(19):6463-6466

[37] Kato T, Gornitzka H, Baceiredo A, Savin A, Bertrand G. On the electronic structure of (phosphine)(silyl) carbenes: Single-crystal X-ray diffraction and ELF analyses. Journal of the American Chemical Society. 2000; **122**(5):998-999

[38] Dyker CA, Lavallo V, Donnadieu B, Bertrand G. Synthesis of an extremely bent acyclic allene (a "carbodicarbene"): A strong donor ligand. Angewandte Chemie International Edition. 2008; **47**(17):3206-3209

[39] Igau A, Baceiredo A, Trinquier G, Bertrand G. [bis (diisopropylamino) phosphino] trimethylsilylcarbene: A stable nucleophilic carbene. Angewandte Chemie International Edition in English. 1989;**28**(5):621-622

[40] Lavallo V, Canac Y, Donnadieu B, Schoeller WW, Bertrand G. Cyclopropenylidenes: From interstellar space to an isolated derivative in the laboratory. Science. 2006;**312**(5774): 722-724

[41] Cavallo L, Cazin CS. N-heterocyclic carbenes: An introductory overview. N-Heterocyclic Carbenes in Transition Metal Catalysis and Organocatalysis. 2010;**32**:1-22

[42] Carter EA, Goddard WA III. Relation between singlet-triplet gaps and bond energies. The Journal of Physical Chemistry. 1986;**90**(6):998-1001

[43] Termaten AT, Schakel M, Ehlers AW, Lutz M, Spek AL, Lammertsma K. N-heterocyclic carbene functionalized iridium phosphinidene complex [Cp*(NHC) Ir= PMes*]: Comparison of phosphinidene, imido, and carbene complexes. Chemistry—A European Journal. 2003;**9**:3577-3582

[44] Lee MT, Hu CH. Density functional study of N-heterocyclic and diaminocarbene complexes: Comparison with phosphines. Organometallics. 2004;**23**(5):976-983

[45] Hu X, Castro-Rodriguez I, Meyer K. Dioxygen activation by a low-valent cobalt complex employing a flexible tripodal N-heterocyclic carbene ligand. Journal of the American Chemical Society. 2004;**126**(41):13464-13473

[46] Cavallo L, Correa A, Costabile C, Jacobsen H. Steric and electronic effects in the bonding of N-heterocyclic ligands to transition metals. Journal of Organometallic Chemistry. 2005;**690** (24–25):5407-5413

[47] Chianese AR, Li X, Janzen MC, Faller JW, Crabtree RH. Rhodium and iridium complexes of N-heterocyclic carbenes via transmetalation: Structure and dynamics. Organometallics. 2003; **22**(8):1663-1667

[48] Dorta R, Stevens ED, Scott NM, Costabile C, Cavallo L, Hoff CD, et al. Steric and electronic properties of N-heterocyclic carbenes (NHC): A detailed study on their interaction with Ni (CO) 4. Journal of the American Chemical Society. 2005;**127**(8):2485-2495

[49] Tolman CA. Steric effects of phosphorus ligands in organometallic chemistry and homogeneous catalysis. Chemical Reviews. 1977;**77**(3):313-348

[50] Huynh HV. Electronic properties of N-heterocyclic carbenes and their experimental determination. Chemical Reviews. 2018;**118**(19):9457-9492

[51] Littke AF, Fu GC. Palladium-catalyzed coupling reactions of aryl chlorides. Angewandte Chemie International Edition. 2002;**41**(22): 4176-4211

[52] Huang J, Schanz HJ, Stevens ED, Nolan SP. Stereoelectronic effects characterizing nucleophilic carbene ligands bound to the Cp* RuCl (Cp*= η5-C5Me5) moiety: A structural and thermochemical investigation. Organometallics. 1999;**18**(12):2370-2375

[53] Dorta R, Stevens ED, Hoff CD, Nolan SP. Stable, three-coordinate Ni $(CO)_2$ (NHC)(NHC= N-heterocyclic carbene) complexes enabling the determination of Ni– NHC bond energies. Journal of the American Chemical Society. 2003;**125**(35): 10490-10491

[54] Back O, Henry-Ellinger M, Martin CD, Martin D, Bertrand G. Angewandte Chemie International Edition. 2013;**52**(10):2939-2943

[55] Biffis A, Gazzola L, Tubaro C, Basato M. Alkyne hydroarylation in ionic liquids catalyzed by palladium (II) complexes. ChemSusChem. 2010;**3**(7): 834-839

[56] Nirmala M, Saranya G, Viswanathamurthi P, Bertani R, Sgarbossa P, Malecki JG. Organonickel complexes encumbering bis-imidazolylidene carbene ligands: Synthesis, X-ray structure and catalytic insights on Buchwald-Hartwig amination reactions. Journal of Organometallic Chemistry. 2017;**831**:1-0

[57] Kreisel KA, Yap GP, Theopold KH. A chelating N-heterocyclic carbene ligand in organochromium chemistry. Organometallics. 2006;**25**(19): 4670-4679

[58] Drent E, Van Broekhoven JA, Doyle MJ. Efficient palladium catalysts for the copolymerization of carbon monoxide with olefins to produce perfectly alternating polyketones. Journal of Organometallic Chemistry. 1991;**417**(1–2):235-251

[59] Farrell K, Müller-Bunz H, Albrecht M. Synthesis, isomerization, and catalytic transfer hydrogenation activity of rhodium (III) complexes containing both chelating dicarbenes and diphosphine ligands. Organometallics. 2015;**34**(24):5723-5733

[60] Farrell K, Melle P, Gossage RA, Müller-Bunz H, Albrecht M. Transfer hydrogenation with abnormal dicarbene rhodium (III) complexes containing ancillary and modular poly-pyridine ligands. Dalton Transactions. 2016; **45**(11):4570-4579

[61] Vivancos A, Albrecht M. Influence of the linker length and coordination mode of (Di) triazolylidene ligands on the structure and catalytic transfer hydrogenation activity of iridium (III) centres. Organometallics. 2017;**36**(8): 1580-1590

[62] Liu Y, Kjaer K, Fredin L, Chabera P, Harlang T, Canton S, et al. A heteroleptic ferrous complex with MesoionicBis (1, 2, 3-triazol-5-ylidene) ligands: Taming the MLCT excited state of iron (II). Chemistry: A European Journal. 2015;**21**(9):3628-3639

[63] Liu Y, Persson P, Sundström V, Wärnmark K. Fe N-heterocyclic carbene complexes as promising photosensitizers. Accounts of Chemical Research. 2016;**49**(8):1477-1485

[64] Desai SP, Mondal M, Choudhury J. Chelating bis-N-heterocyclic carbene–palladium (II) complexes for oxidative arene C–H functionalization. Organometallics. 2015;**34**(12):2731-2736

[65] McCall AS, Wang H, Desper JM, Kraft S. Bis-N-heterocyclic carbene palladium (IV) tetrachloride complexes: Synthesis, reactivity, and mechanisms

of direct chlorinations and oxidations of organic substrates. Journal of the American Chemical Society. 2011; **133**(6):1832-1848

[66] Meyer D, Ahrens S, Strassner T. Platinum (IV) complexes with chelating N-heterocyclic carbene ligands. Organometallics. 2010;**15**(29):3392-3396

[67] Prokopchuk EM, Puddephatt RJ. Hydrido (methyl) carbene complex of platinum (IV). Organometallics. 2003; **22**(3):563-566

[68] Lindner R, Wagner C, Steinborn D. Synthesis of Trimethylplatinum (IV) complexes with N, N-and N, O-heterocyclic carbene ligands and their reductive C– C elimination reactions. Journal of the American Chemical Society. 2009;**131**(25):8861-8874

[69] Arnold PL, Sanford MS, Pearson SM. Chelating N-heterocyclic carbene alkoxide as a supporting ligand for PdII/IV C–H bond functionalization catalysis. Journal of the American Chemical Society. 2009;**131**(39): 13912-13913

[70] Aznarez F, Iglesias M, Hepp A, Veit B, Miguel PJ, Oro LA, et al. Iridium (III) complexes bearing chelating bis-NHC ligands and their application in the catalytic reduction of imines. European Journal of Inorganic Chemistry. 2016; **28**(2016):4598-4603

[71] Galan BR, Gembicky M, Dominiak PM, Keister JB, Diver ST. Carbon monoxide-promoted carbene insertion into the aryl substituent of an N-heterocyclic carbene ligand: Buchner reaction in a ruthenium carbene complex. Journal of the American Chemical Society. 2005;**127**(45): 15702-15703

[72] Allen DP, Crudden CM, Calhoun LA, Wang R. Irreversible cleavage of a carbene–rhodium bond in Rh-N-heterocyclic carbene complexes: Implications for catalysis. Journal of Organometallic Chemistry. 2004; **689**(20):3203-3209

[73] Danopoulos AA, Tsoureas N, Green JC, Hursthouse MB. Migratory insertion in N-heterocyclic carbene complexes of palladium; an experimental and DFT study. Chemical Communications. 2003;**6**:756-757

[74] Jazzar RF, Macgregor SA, Mahon MF, Richards SP, Whittlesey MK. C– C and C– H bond activation reactions in N-heterocyclic carbene complexes of ruthenium. Journal of the American Chemical Society. 2002;**124**(18):4944-4945

[75] Becker E, Stingl V, Dazinger G, Puchberger M, Mereiter K, Kirchner K. Facile migratory insertion of a N-heterocyclic carbene into a ruthenium–carbon double bond: A new type of reaction of a NHC ligand. Journal of the American Chemical Society. 2006; **128**(20):6572-6573

[76] McGuinness DS, Saendig N, Yates BF, Cavell KJ. Kinetic and density functional studies on alkyl-carbene elimination from pdii heterocyclic carbene complexes: A new type of reductive elimination with clear implications for catalysis. Journal of the American Chemical Society. 2001; **123**(17):4029-4040

[77] Simms RW, Drewitt MJ, Baird MC. Equilibration between a phosphine–cobalt complex and an analogous complex containing an N-heterocyclic carbene: The thermodynamics of a phosphine– carbene exchange reaction. Organometallics. 2002;**21**(14): 2958-2963

[78] Titcomb LR, Caddick S, Cloke FGN, Wilson DJ, McKerrecher D. Unexpected reactivity of two-coordinate palladium–carbene complexes; synthetic and catalytic implications. Chemical Communications. 2001;**15**:1388-1389

[79] Liske A, Verlinden K, Buhl H, Schaper K, Ganter C. Determining the π-acceptor properties of N-heterocyclic carbenes by measuring the 77Se NMR chemical shifts of their selenium adducts. Organometallics. 2013;**32**(19):5269-5272

[80] Boehme C, Frenking G. Electronic structure of stable carbenes, silylenes, and germylenes. Journal of the American Chemical Society. 1996; **118**(8):2039-2046

[81] Arduengo AJ III, Dias HR, Harlow RL, Kline M. Electronic stabilization of nucleophilic carbenes. Journal of the American Chemical Society. 1992;**114**(14):5530-5534

[82] Denk MK, Thadani A, Hatano K, Lough AJ. Steric stabilization of nucleophilic carbenes. Angewandte Chemie International Edition in English. 1997;**36**(23):2607-2609

[83] Hahn FE, Paas M, Le Van D, Lügger T. Simple access to unsymmetrically substituted, saturated N-heterocyclic carbenes. Angewandte Chemie International Edition. 2003; **42**(42):5243-5246

[84] Hahn FE, Paas M, Le Van D, Fröhlich R. Spirocyclicdiaminocarbenes: Synthesis, coordination chemistry, and investigation of their dimerization behavior. Chemistry—A European Journal. 2005;**11**(17):5080-5085

[85] Glorius F, Altenhoff G, Goddard R, Lehmann C. Oxazolines as chiral building blocks for imidazolium salts and N-heterocyclic carbene ligands. Chemical Communications. 2002;**22**: 2704-2705

[86] Kuhn N, Kratz T. Synthesis of imidazol-2-ylidenes by reduction of imidazole-2 (3H)-thiones. Synthesis. 1993;**1993**(06):561-562

[87] Scholl M, Ding S, Lee CW, Grubbs RH. Synthesis and activity of a new generation of ruthenium-based olefin metathesis catalysts coordinated with 1, 3-dimesityl-4, 5-dihydroimidazol-2-ylidene ligands. Organic Letters. 1999;**1**(6):953-956

[88] Wanzlick HW, Kleiner HJ. Nucleophilic carbene chemistry Representation of bis[1.3-diphenyl-imidazolidinylidene-(2)]. Applied Chemistry. 1961;**73**(14):493

[89] Nyce GW, Csihony S, Waymouth RM, Hedrick JL. A general and versatile approach to thermally generated N-heterocyclic carbenes. Chemistry—A European Journal. 2004; **10**(16):4073-4079

[90] Blum AP, Ritter T, Grubbs RH. Synthesis of N-heterocyclic carbene-containing metal complexes from 2-(pentafluorophenyl) imidazolidines. Organometallics. 2007;**26**(8): 2122-2124

[91] Korotkikh NI, Raenko GF, Pekhtereva TM, Shvaika OP, Cowley AH, Jones JN. Stable carbenes. Synthesis and properties of benzimidazol-2-ylidenes. Russian Journal of Organic Chemistry. 2006; **42**(12):1822-1833

[92] Otto M, Conejero S, Canac Y, Romanenko VD, Rudzevitch V, Bertrand G. Journal of the American Chemical Society. 2004;**126**: 1016-1017

[93] Arduengo AJ III, Krafczyk R, Schmutzler R, Craig HA̧, Goerlich JR, Marshall WJ, et al. Imidazolylidenes, imidazolinylidenes and imidazolidines. Tetrahedron. 1999;**55**(51):14523-14534

[94] Herrmann WA, Köcher C, Gooßen LJ, Artus GR. Heterocyclic carbenes: A high-yielding synthesis of novel, functionalized N-heterocyclic carbenes in liquid ammonia. Chemistry–A European Journal. 1996;**2**(12): 1627-1636

[95] Glorius F, Altenhoff G, Goddard R, Lehmann C. Oxazolines as chiral building blocks for imidazolium salts and N-heterocyclic carbene ligands. Chemical Communications. 2002;**22**: 2704-2705

[96] Wang AE, Xie JH, Wang LX, Zhou QL. Triaryl phosphine-functionalized N-heterocyclic carbene ligands for Heck reaction. Tetrahedron. 2005;**61**(1):259-266

[97] Perry MC, Cui X, Powell MT, Hou DR, Reibenspies JH, Burgess K. Optically active iridium imidazole-2-ylidene-oxazoline complexes: Preparation and use in asymmetric hydrogenation of aryl alkenes. Journal of the American Chemical Society. 2003;**125**(1):113-123

[98] César V, Bellemin-Laponnaz S, Gade LH. Direct coupling of oxazolines and N-heterocyclic carbenes: A modular approach to a new class of c– n donor ligands for homogeneous catalysis. Organometallics. 2002;**21**(24):5204-5208

[99] Loch JA, Albrecht M, Peris E, Mata J, Faller JW, Crabtree RH. Palladium complexes with tridentate pincer bis-carbene ligands as efficient catalysts for C–C coupling. Organometallics. 2002;**21**(4):700-706

[100] Yang C, Lee HM, Nolan SP. Highly efficient Heck reactions of aryl bromides with n-butyl acrylate mediated by a palladium/phosphine–imidazolium salt system. Organic Letters. 2001;**3**(10):1511-1514

[101] Gardiner MG, Herrmann WA, Reisinger CP, Schwarz J, Spiegler M. Dicationic chelating N-heterocyclic carbene complexes of palladium: New catalysts for the copolymerisation of C_2H_4 and CO. Journal of Organometallic Chemistry. 1999;**572**(2):239-247

[102] Herrmann WA, Goossen LJ, Spiegler M. Chiral oxazoline/imidazoline-2-ylidene complexes. Organometallics. 1998;**17**(11):2162-2168

[103] Herrmann WA, Gooβen LJ, Spiegler M. Functionalized imidazoline-2-ylidene complexes of rhodium and palladium. Journal of Organometallic Chemistry. 1997;**547**(2):357-366

[104] Fürstner A, Alcarazo M, César V, Lehmann CW. Convenient, scalable and flexible method for the preparation of imidazolium salts with previously inaccessible substitution patterns. Chemical Communications. 2006;**20**: 2176-2178

Chapter 4

Basic Information about Carbenes

Nuriye Tuna Subasi

Abstract

In organic chemistry, good knowledge of the chemistry of reactive intermediates is very important in understanding organic chemistry, which has a systematic like mathematics. Having this information, it is possible to predict the reactions and the products to be formed on the reactive intermediates. Otherwise, it becomes impossible to learn organic chemistry, which has a very wide scope. This chapter of the book will be a guide that summarizes the essential information about carbene, one of the important reactive intermediates. It is planned to include the following sub-headings in this chapter: Carbene definition and properties, Nomenclature of carbenes, Structure and reactivity of carbenes, Carbene synthesis, Carbene reactions.

Keywords: carbenes, triplet and singlet carbenes, diazo compounds, carbene synthesis, carbene reactions

1. Introduction

Carbenes are neutral, divalent (two atoms attached to the central carbon atom) and highly reactive carbon intermediates generally written as "R_2C:". Carbenes, which have six electrons in their outer orbital, have a nonbonding pair of electrons and their formal charge is zero. They are short-lived and exhibit extreme reactivity because of electron deficiency.

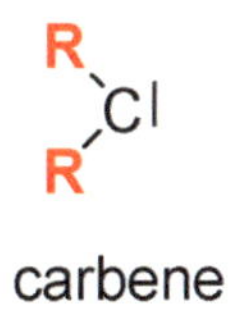

R = C, N, O, S, F, Cl, Br, I .....

The simplest carbene is the CH_2 compound and is called methylene, the term being first introduced during 19th century [1]. As early as 1835, Dumas [2] reported his attempts to prepare the parent carbene (CH_2) by dehydration of methanol. Actually at that time, the quadrivalency of carbon atom was not exactly established and for this reason many ancient chemist believed that methylene would be a stable compound, and various experiments were carried out to produce it. At the end of the 19th century, Curtius [3] and Staudinger [4] showed that carbenes, produced from diazo compounds or ketenes, were highly reactive species. In addition to this, carbenes became popular as transient species [5] in the 1940–1950s, when Doering [6] discovered the cyclopropanation reaction. Then the word carbene was first used by Woodward, Doering and Winstein at a meeting of the American Chemical Society in 1951.

IntechOpen

Carbenes are generally classified as triplet and singlet. As we will explain in detail later in singlet carbenes the two electrons are paired and the molecule is diamagnetic, while for triplet carbenes the two electrons are unpaired and the molecule is paramagnetic. Carbenes can be either linear or bent.

2. Nomenclature of carbenes

Today, the term carbene is used for divalent carbon species, and they are commonly referred to together with substituents. Substituents are given first, and the word carbene is added to the end.

If the divalent carbon atom is in a ring or the carbene electrons are on a carbon–carbon double bond carbon atom, these carbenes are named by the suffix *-yilidene* [7].

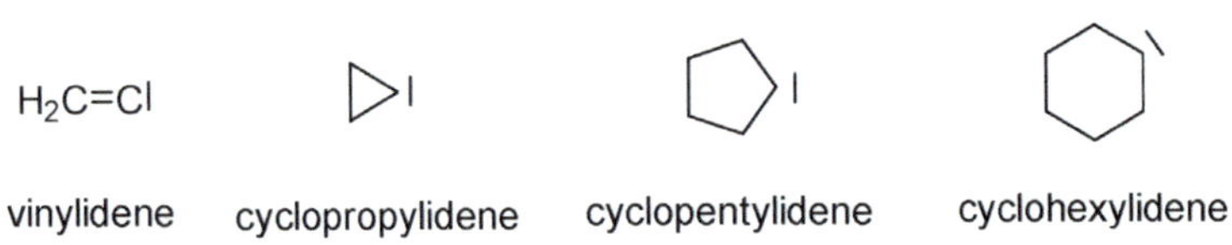

3. Structure and reactivity

Generally, carbenes have two bonding electrons (both in sp^2-orbitals) and two non-bonding electrons. There are two classes of carbenes called singlet or triplet carbene depending on whether the non-bonding electrons are in the same or different orbitals, respectively as shown in **Figure 1**. Carbenes usually contain sp^2 hybridized carbon atoms according to the valance bond theory. Two of the three sp^2-hybrid orbitals bond with their carbene substituents by covalent bonds and two vacant orbitals remain, consisting of the sp^2-hybrid orbital and the p orbital. Two non-bonding carbene electrons must be placed in these vacant orbitals. If two electrons are placed in the same orbital, this carbene is called a singlet carbene, since the electron spins will be in the opposite directions. When electrons are placed in different orbitals, parallel spin will be preferred according to Hund's law, and the formed carbene is called triplet carbene [7–11].

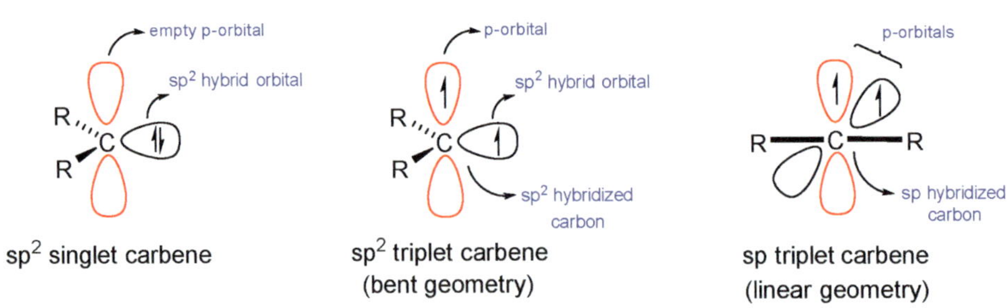

Figure 1.
Electronic structures of singlet and triplet carbene.

Triplet carbenes have an angular structure, as well as a linear structure. In carbene, which has a linear structure, the carbon atom to which the substituents are attached makes the sp. hybridization and there are two vacant p orbitals in the molecule. Since the energy levels of these orbitals will be equal, according to Hund's law, carbene electrons are placed in these orbitals one by one and carbene gains the triplet property [7–11].

Generally in most of the organic compounds, the singlet state is more stable than the triplet state. As a result the ground states of these molecules are singlet. In these molecules triplet state occurs only as excited or high energy level. On the contrary, studies show that carbenes usually have an angular structure and their electronic structure is triplet in the ground state because triplet carbenes have lower energy so they are more stable than singlet carbenes. Energy difference between singlet and triplet carbene is 8 kcal/mol.

The nature of substituents affects the electronic properties of carbenes. If the substituents attached to the carbene carbon are electron withdrawing groups, the carbene prefers the singlet structure. Electron withdrawing groups inductively stabilize the σ-orbital attached to the carbene carbon, increasing the energy difference between the σ- and π-orbital. Thus, electrons place in the σ-orbital, leaving the π-orbital empty. Conversely, if the substituents attached to the carbene carbon are electron donating groups via the σ-bond, the carbene prefers the triplet structure [7–11].

In addition, if the atoms attached to the carbene carbon have non-bonding electron pairs (nitrogen, oxygen, sulfur, halogen, etc.), these atoms easily donate their electrons to the vacant p-orbital of carbene. Thus, this π-donor atoms stabilize the singlet state by resonance structure and then carbenes prefer the singlet configuration in the ground state. For example, dichlorocarbene is singlet in the ground state. As shown in **Figure 2**, electrons on the chlorine atom are conjugated with the carbene atom, increasing the stability of the carbene [7–12].

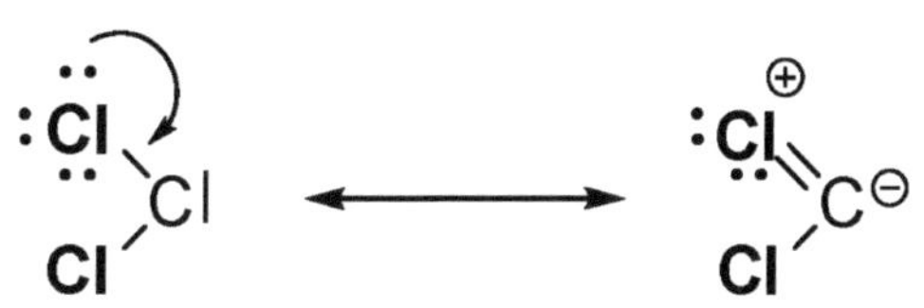

Figure 2.
Stabilization of dicholorocarbene singlet state by π-donation (mesomeric effect).

Unstabilized carbenes usually have triplet ground states due to their stability, [9] but lone pair donating substituents can reverse this situation. As a consequence, electronic substituent effects, typically π-donation (such as $-NR_2$, -OR, -SR, -F, -Cl, -Br, -I) and π -acceptance (such as -COR, -SOR, $-SO_2R$, -NO, $-NO_2$) [13], as well as hyperconjugation (by alkyl groups) [14] and electronegative substituents [15] mainly stabilize the singlet state. Moreover, electropositive substituents with at least one atom having non-bonding electron pair give singlet carbene [16].

The nature of the substituents affect the chemical reactivity of carbenes as well as their electronic structure. Since carbenes are electron deficient intermediates (the carbon atom having only six electron in its outer shell), they show electrophilic behavior in their reactions. Naturally, when electron withdrawing groups are attached to these carbenes, the electrophilicity of the carbene increases. Nevertheless, if very strong π-donor substituents are attached to the carbene intermediate then it behaves as nucleophile in its reactions. For example, diaminocarbenes (**Figure 3**) are nucleophilic singlet carbenes because of π-donation of substituent [1, 7].

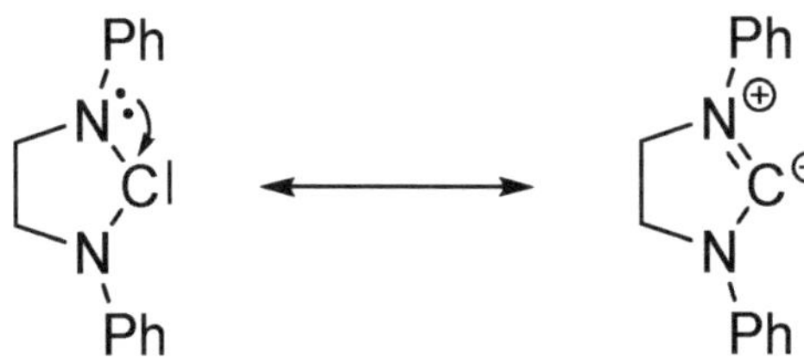

Figure 3.
Example of diaminocarbenes (nucleophilic singlet carbene).

Another example of nucleophilic singlet carbenes is the cycloheptatrienylidene molecule. Since the vacant p-orbital located on the carbene atom participates in the delocalization of π-system of the seventh ring, the carbene electrons have to be placed in the sp^2 hybrid orbital on the ring plane. Therefore, cycloheptatrienylidene is a singlet and nucleophilic carbene [1, 7].

It is not possible to observe carbenes under normal conditions. However, by using the matrix isolation method carbenes formed by photolysis of diazo compounds at 4–77 K in frozen argon or cyclohexane can be observed by IR or ESR (Electron Spin Resonance) spectroscopy [1]. Since ESR is a spectroscopic method based on electron spin, triplet carbenes can be observed with this method. Triplet carbenes, which act like diradicals, can be observed with ESR due to these properties. ESR not only clearly defines triplet carbenes, but also gives information about the molecular and electronic structures of carbenes [17–20]. For example, with this method, it was determined that triplet methylene and diphenylcarbene were angular and these angles were 136° and 142°, respectively.

Carbenes can be stabilized by steric or electronic effects [9]. As a result of the studies carried out in the light of this idea, some stable carbene molecules can be isolated at room temperature as shown in **Figure 4**. In 1988, Bertrand [21] et al reported the synthesis of the (phosphino)(silyl)carbene (first isolated carbene) **1** which can remain stable for weeks at room temperature. This compound, obtained by the decomposition of diazo compounds, was isolated by vacumm distillation (10^{-2} Torr) at 75–80°C as a red oily material in 80% yield. It has all the typical reactivity associated with "classic" carbenes [22]. A few years later the first crystalline carbene was reported by Arduengo and co-workers [23]. This discovery catalyzed research activities of carbene. The 1,3-diadamantylimidazol-2-ylidene **2** can easily synthesized in 96% yield under an inert atmosphere even at room temperature. Bond angle of 1,3-diadamantylimidazol-2-ylidene, a colorless and crystalline compound with a melting point of 240°C, was determined as 102° by X-ray analysis. There are two main factors that stabilize this molecule. Nitrogen atoms attached to the carbene atom stabilize the carbene electronically, while adamantane groups attached to the nitrogen atoms make it sterically stable.

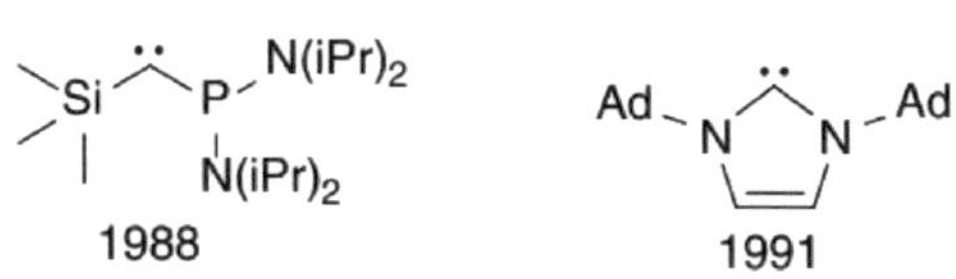

1 **2**

Figure 4.
The first isolated carbenes. iPr = isopropyl, Ad = adamantyl.

4. Generation of carbenes

The synthesis of carbene molecules is generally based on elimination and fragmentation reactions. Carbenes are formed as intermediate products when groups attached to the carbon atom are broken as a result of photolysis, thermolysis or reaction with metals.

4.1 Fragmentation reactions

Diazo compounds, tosylhydrazones, ketenes, three-membered carbon rings, strained alkenes and heterocyclic compounds are generally used as carbene precursors in fragmentation reactions.

4.1.1 Diazo compounds

Diazo compounds (RR'C=N2), constitute a principal class of carbene precursors, known since the first preparation of ethyl diazoacetate by Curtius in 1883 [24]. Diazo compounds, which have 1,3-dipolar structures (**Figure 5a**), are generally converted to related carbenes by easily removing nitrogen gas when heated or photolyzed in aprotic solvents as shown in **Figure 5b**. This conversion has been proven by spectroscopic methods. The electronic spectrum of methylene produced in the gas phase by flash photolysis of diazomethane was recorded by Herzberg and Shoosmith [25]. Various metal complexes are used as catalyst for carbene formation to occur at low temperatures. Since diazo compounds are generally unstable and easily degradable compounds, they must be synthesized before each reaction.

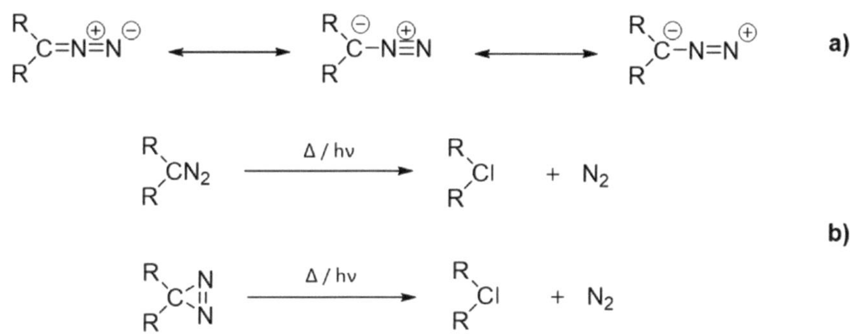

Figure 5.
(a) Resonance structure of 1,3-dipolar diazo compound. (b) Generation of carbene from diazo compounds.

Diazo compounds, which have a unique reactivity due to their 1,3-dipole and ylide structures, are useful synthetic products in organic synthesis. However, their large-scale use has been avoided due to their toxicity and unpredictable explosive behavior [26].

4.1.2 Tosylhydrazones

Where the diazo compound is somehow unstable and dangerous to use, it is usually better to use a diazo precursor. The simplest and most common compounds used for this are hydrazones. Ketones and aldehydes easily react with hydrazine to form hydrazone compounds. The oxidation of hydrazones with metal salts such as Ag_2O, HgO, MnO_2, $Pb(OAc)_4$ gives the diazo compound (**Figure 6**).

The most widely known and used carbene precursors are tosylhydrazones which are prepared from the reaction of aldehydes and ketones with p-toluenesulphonyl hydrazide.

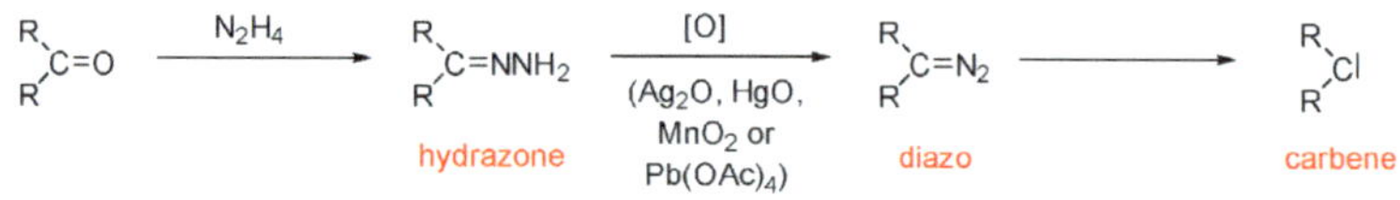

Figure 6.
Generation of carbenes from hydrazones compound.

p-Toluenesulpnonylhydrazones (tosylhydrazones) of aldehydes and ketones undergo base-catalyzed thermal decomposition with loss of p-toluenesulfinate to give intermediate diazo compounds [27]. This method is called the Bamford-Stevens reaction (**Figure** 7).

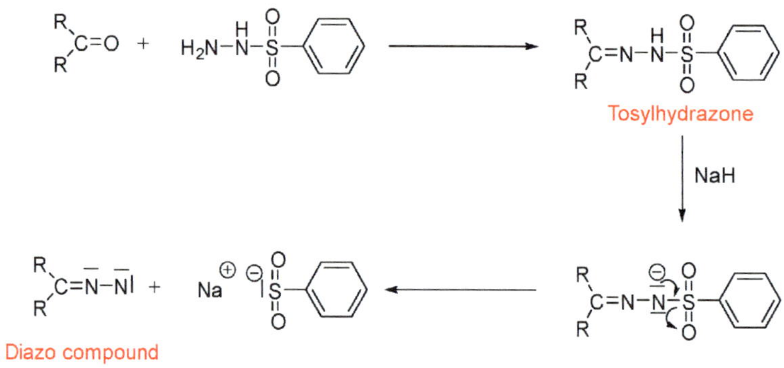

Figure 7.
Bamford-Stevens reaction.

When the N-H proton in the tosylhydrazone molecule is removed with a base such as NaH or $NaOCH_3$, the formed anion is stabilized by the tosyl group. Therefore, the resulting salt can be isolated and stored for a long time. This salt can be converted to carbene by heating at any time in situ. After the nitrogen gas is separated, alkene formation is observed as a result of 1,2-hydrogen shift. When the alkene formation mechanism was examined comprehensively, it was found that the products formed may vary depending on the reaction conditions [28]. In the Bamford-Stevens reaction, if the decomposition of the tosylhydrazone salt is done in aprotic solvents, carbene is formed by the removal of nitrogen gas. The resulting carben usually forms an alkene as a result of 1,2-hydrogen shift. However, if the salt is decomposed in protic solvents, the diazonium salt is formed first. Diazonium salts, especially the aliphatic ones, are not stable, they turn into carbenium ions by removing nitrogen gas.

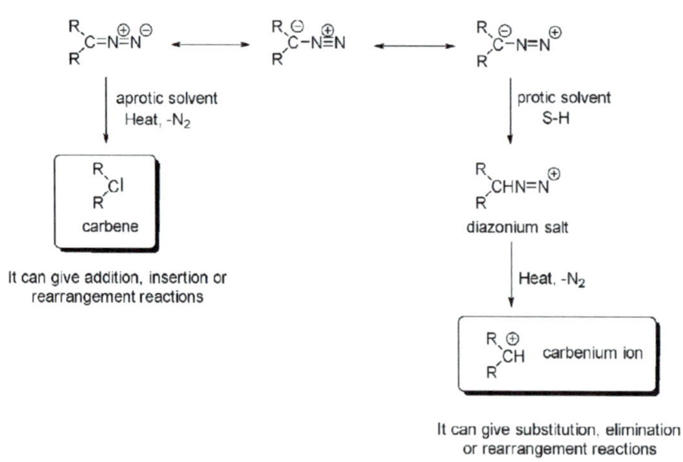

4.1.3 Ketenes

Ketenes can eliminate CO molecule on thermolysis or photolysis to generate carben. Since ketenes are not readily available precursor and polymerize under the reaction conditions, they are not widely used. Ketene has been used extensively to generate CH_2.

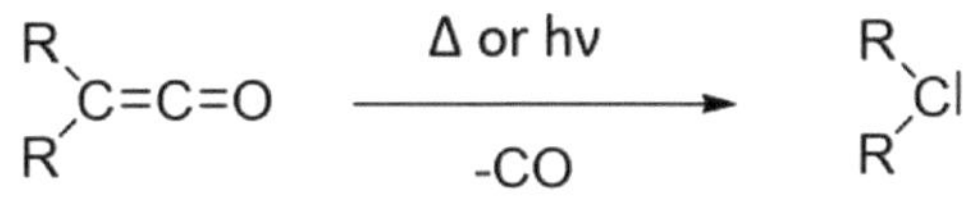

4.1.4 Three-membered carbon ring

Three-membered rings that have a high ground state energy due to steric strain often decompose to give carbene intermediates by heating or irradiation. For example; photolysis of 1,1-dichloro-2-phenylcyclopropane apparently gives CCl_2 and photolytic decomposition of oxiranes yield arylcarbenes and related species as shown in **Figure 8**.

Substituted phenyloxiranes provide convenient precursors for substituted arylcarbenes, such as diphenylcarbene [29], phenylmethylcarbene [30], cyanophenylcarbene [31, 32], and methoxycarbonylphenylcarbene [33]. Selective cleavage was observed with unsymmetrical precursors, the thermodynamically more stable isomer appears to be favored.

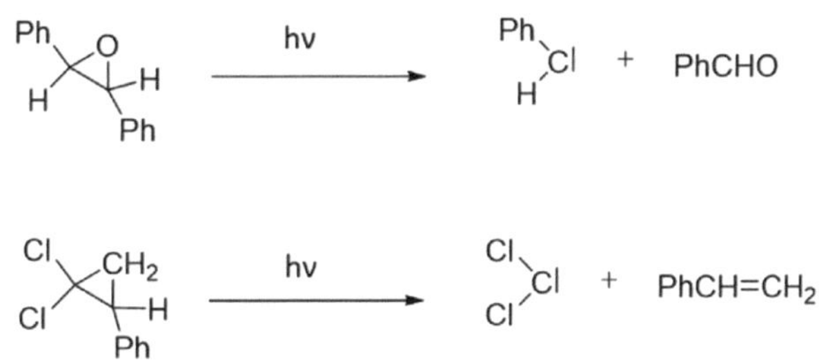

Figure 8.
Generation of carbene from cyclopropanes and epoxides.

The most useful carbene precursors of the three-membered ring are diazirines. Diazirines, the cyclic isomers of diazoalkanes, also decompose under the influence of heat and light to give carbenes [34–37]. As shown in **Figure 9**, they are prepared from ketones by reaction with ammonia and chloramine followed by oxidation of the resulting diaziridine. This method is quite widely used especially for halocarbenes.

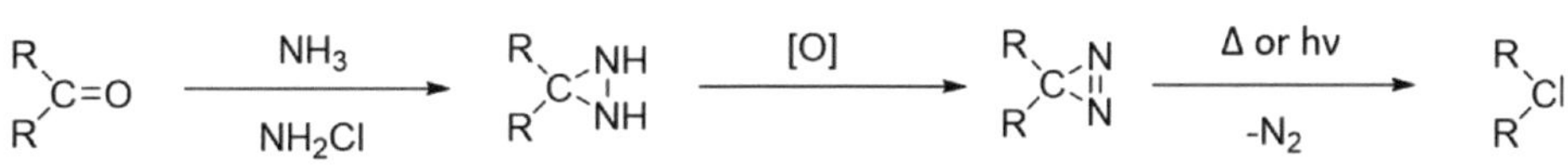

Figure 9.
Preparation and decomposition of diazirines.

4.1.5 Strained alkenes

If the alkene is extremely sterically hindered then the π-bond is weakend due to the substantially reduced p–p overlap and distortion from planarity. As a result, the ground state energy is raised and then dissociation to carbenes become possible by heating. The well-known example of this process is the reversible dissociation of tetranaphth-1-ylethene into bis(naphth-1-yl) carbene at 250°C [38] (**Figure 10**).

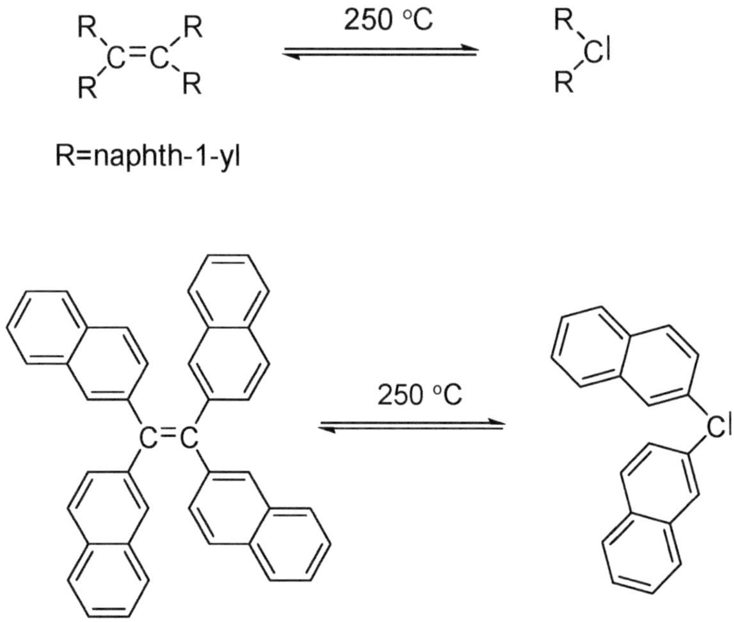

Figure 10.
Thermal dissociation of tetranaphth-1-ylethene.

4.1.6 Heterocycles

Various five-membered heterocyclic compounds decompose to give carbenes by heating or irradiation (**Figure 11**). The decomposition of 1,5-dihydro-1,3,4-oxadiazoles give carbenes at about 80°C with loss of nitrogen followed by the carbonyl fragment.

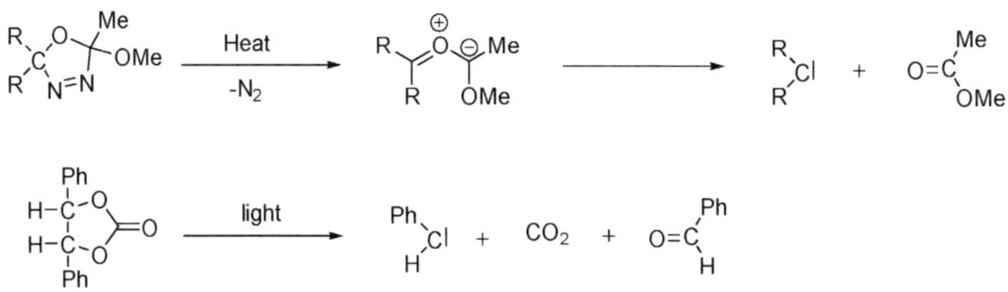

Figure 11.
Fragmentation of five-membered heterocycles to carbenes.

4.2 α-Elimination

The hydrolysis of chloroform in basic medium was probably the first reaction in which intermediate carbene formation was suggested by Geuther [39]. The investigation the mechanism of this reaction by Hine and co-workers initiated the modern era of carben chemistry in the early 1950s [40–42]. In the basic environment, the acidic proton of chloroform separates and trichloromethyl anion is formed.

Dichlorocarbene is formed as a result of the removal of the chloride anion from the carbanion. By hydrolysis of dichlorocarbene in aqueous media, carbon monoxide is formed (**Figure 12**). These types of reactions are called α-elimination reactions because the hydrogen and chlorine released in the formation of carbene are attached to the same carbon atom.

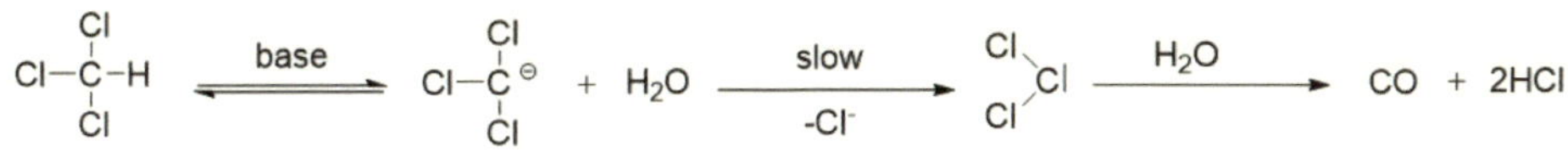

Figure 12.
α-Elimination reactions.

In experiments with aprotic solvents and strong bases, it was determined that carbene was formed and added to the double bond electrons to form cyclopropane derivatives as shown in **Figure 13**. This reaction is the best method for cyclopropane synthesis in organic chemistry.

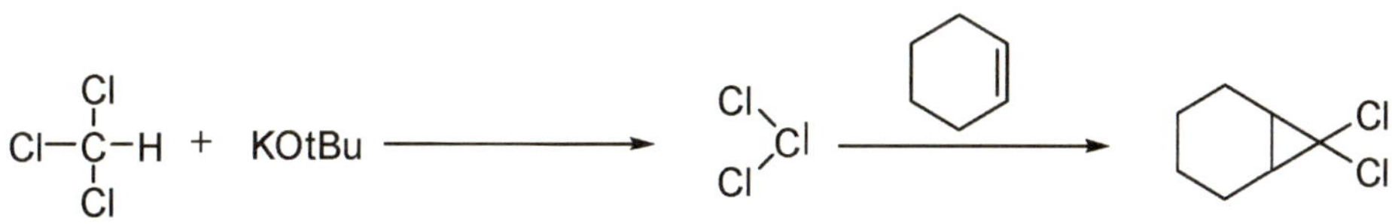

Figure 13.
Cyclopropane synthesis.

Before the reaction, the solvent used must be dried very well. Because the water in the environment can easily react with the carbene formed, and it also reacts with the base in the environment and completely prevents the formation of carbenes. However, in Makosza, a study carried out in a two-phase system, showed that carbenes were formed in the presence of phase transfer catalysts in aqueous medium and added to double bonds (**Figure 14**) [43].

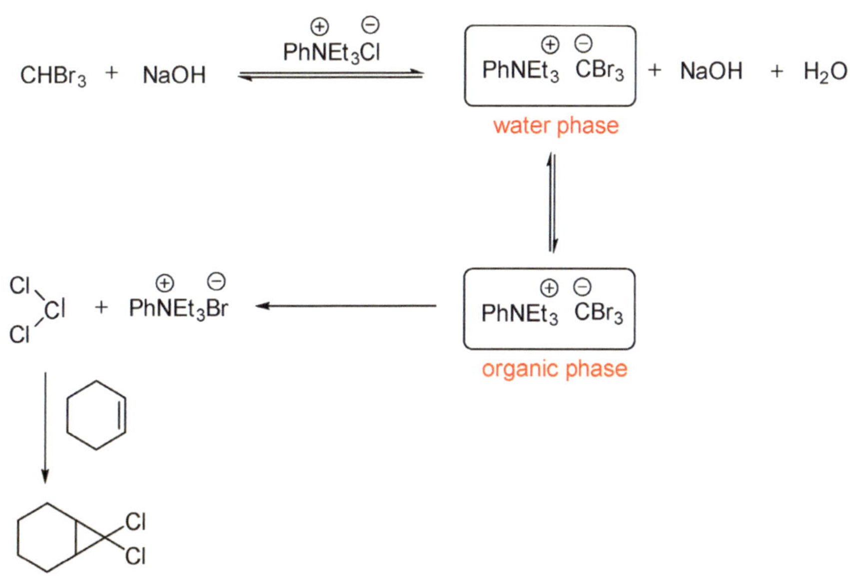

Figure 14.
Carbene formation in a two-phase system.

Since the reaction is between two phases, effective mixing is very important for this reaction. Although the reaction is carried out in the presence of water, the carbene is generated and reacts in the organic phase.

4.2.1 Simmon-Smith reaction

It is not possible to synthesize methylene carbene by the α-elimination method. However, there are methods that form methylene carbene adducts. The most commonly used method is the reaction of alkenes with diiodomethane in the presence of zinc. As a result of the reaction, cyclopropane compounds are formed (**Figure 15**) [44, 45].

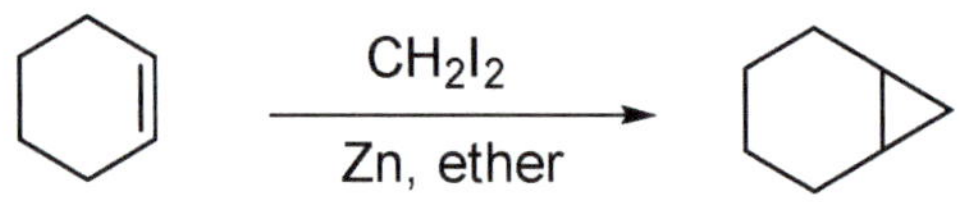

Figure 15.
Simmon-Smith reaction.

During this reaction, methylene carbene is not formed in the free form. First, zinc and diiodomethane react to form a carbenoid intermediate, which acts as a carbene. Later, since this product is unstable, it transforms into zinc iodide by transferring the methylene group to the double bond as shown in **Figure 16** [46].

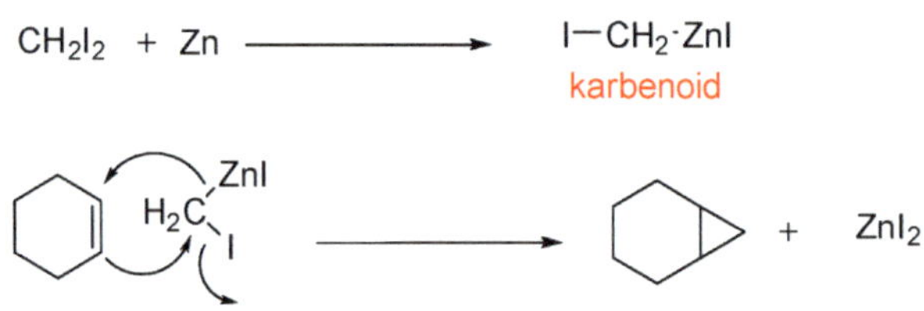

Figure 16.
Mechanism of Simmon-Smith reaction.

5. Carbene reactions

Singlet and triplet carbene exhibit different reactivity. Singlet carbenes generally participate in reactions as either electrophiles or nucleophiles. Singlet carbenes which have unfilled p-orbitals should be electrophilic. Triplet carbenes can be considered as diradicals and participate in stepwise radical additions. Triplet carbenes must pass through an intermediate with two unpaired electrons, while single carbenes can react in a single concerted step. Because carbenes have two modes of reactivity, singlet methylene gives stereospecific reactions where as triplet methylene gives stereoselective reactions [47].

Carbenes are highly reactive intermediate due to electron deficiency. Carbenes react instantly in various ways in the environment where they are formed. Carbene reactions are classified under four main groups.

5.1 Cycloaddition reaction of carbenes

Since carbenes generally react electrophilically, they give a [2 + 2] cycloaddition reaction with double bonds to form cyclopropane compounds (**Figure 17**). This

Figure 17.
Cycloaddition reaction of carbenes.

method, the most characteristic reaction of carbene intermediates, has now been widely used as a synthetic route to cyclopropane since 1954 [6].

According to the theory proposed by Skell and Woodworth [47], singlet carbenes are added the double bonds simultaneously in a single step. The opening of double bond electrons and the formation of new carbon–carbon bonds occur simultaneously. On the other hand, triplet carbenes are not added in double bonds in a single step because this addition is forbidden by orbital symmetry. Triplet carbenes form cyclopropane compounds by adding double bonds as a result of a multi-step reaction.

Since carbenes are generally electrophilic compounds, they prefer double bonds where the electron density is high when adding systems that contain more than one double bond. In particular, singlet carbenes (dihalocarbenes) show more regioselective properties as they are more stable. For example, in the isotetralin compound, which has two different double bonds, dihalocarbene selectively adds the central double bond, although it has a steric hindrance. However, the more reactive ethoxycarbonylcarbene cannot act selectively due to its shorter lifetime and adds the double bonds located on the outer part of the molecule as shown below.

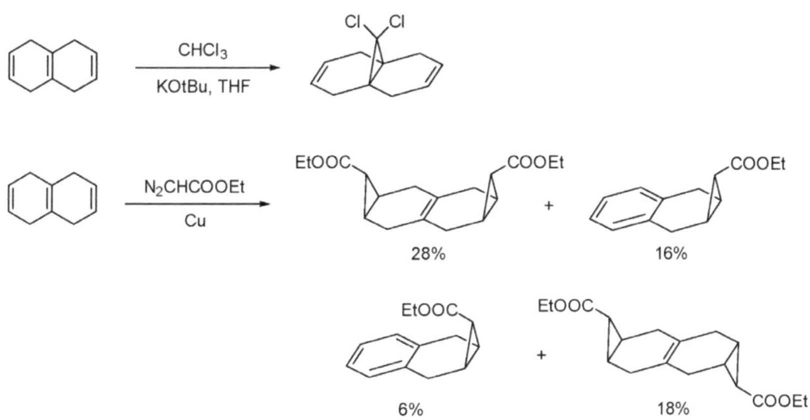

5.2 Dimerization reaction of carbenes

When carbenes cannot find a reagent that can react in the environment in which they are formed, they dimerize to fill the electron gap in their outer orbitals and turn into olefins (**Figure 18**).

Figure 18.
Dimerization reaction of carbenes.

5.3 Insertion reaction of carbenes

Another characteristic reaction that carbenes give to become stable is that carbenes insert between C-H or C-C bonds (**Figure 19**).

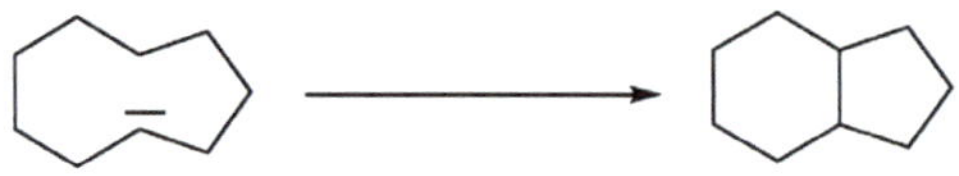

Figure 19.
Examples of carbene insertion reaction.

Methylene, when there is no double bond to react it is not selective and insert into a random C-H bond in the liquid phase. Methylene acts selectively in the gas phase and preferentially inserts into tertiary C-H bonds as shown in **Figure 20**.

Figure 20.
Insertion of methylene.

The electronic structure of carbene plays an important role in the insertion reactions (**Figure 21**). Singlet carbenes generally insert into bonds in a single step with retention of configuration. The situation is different with triplet carbenes. Since triplet carbenes act as radicals, they first abstract hydrogen from the C-H bond and form two new radicals. With the combination of these radicals, a new C-C bond is formed. Meanwhile, as the radical configuration undergoes isomerization, a racemic mixture is formed as a result of the reaction. In the insertion of triplet carbenes, tertiary carbon-hydrogen bonds are primarily preferred because they form stable radicals.

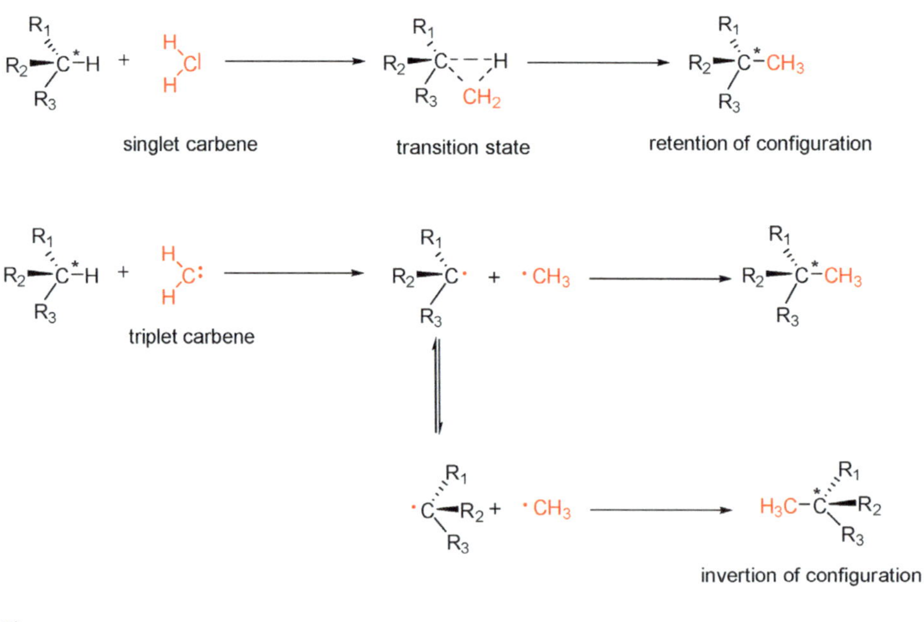

Figure 21.
Mechanism of singlet and triplet carbene in the insertion reaction.

5.4 Rearrangement of carbenes

Since carbenes are electron-deficient intermediates with an empty p-orbital, an atom or group on the neighboring carbon migrates to the electron-deficient center and undergoes easy rearrangement with simultaneous formation of a new C=C bond. Usually this rearrangement is called 1,2-shift, and it usually involves the migration of a hydrogen atom, since the order of migrating ability is: H> > aryl>alkyl. This hydrogen shift is also considered as the intramolecular addition of the carbene to the adjacent C-H bond.

The best known carbene rearrangement is the rearrangement of diazoketones to ketenes. In this reaction, called Wolff rearrangement, α-diazoketo compounds primarily form α-ketocarbene by removing nitrogen gas thermally, photochemically, or in the presence of various metals. When the -R group electrons attached to the carbonyl group migrate to the carbene center, the carbene electrons form a double bond with the carbonyl carbon and ketene is formed. Later, ketenes react with water and turn into carboxylic acids [48]. According to the experiments, the Wolff rearrangement proceeds according to a concerted mechanism in which the migration occurs at the same time as the leaving group and the leaving nitrogen group and the migrating alkyl group should be in antiperiplanar conformation as shown in **Figure 22**.

Figure 22.
Wolff rearrangement.

When Wolff rearrangement is applied to cyclic systems, a ring reduction reaction occurs if the carbene located at the α-position to the carbonyl group is formed in a ring (**Figure 23**).

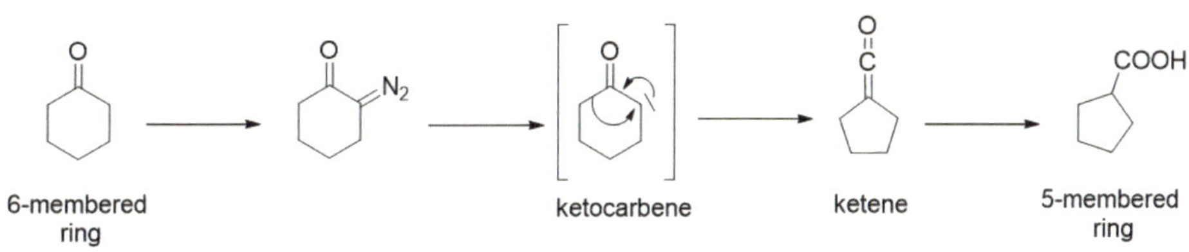

Figure 23.
Ring reduction reaction in the cyclic system.

If the carbene is attached to the cyclic structure as a substituent, then a ring expansion reaction occurs (**Figure 24**).

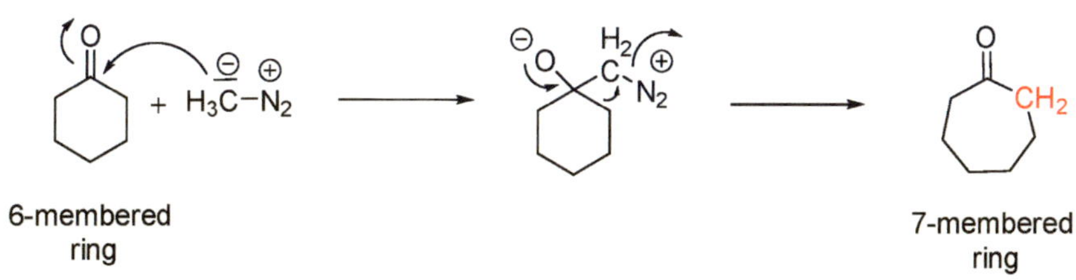

Figure 24.
Ring expansion reaction in the cyclic system.

6. Use of carbenes in organic synthesis

6.1 Example 1

Preparation of tetracyclic cyclopropapyrroloindole, an analogue of the mitosene antitumor agent, involves an intramolecular carbene addition to an alkene (**Figure 25**).

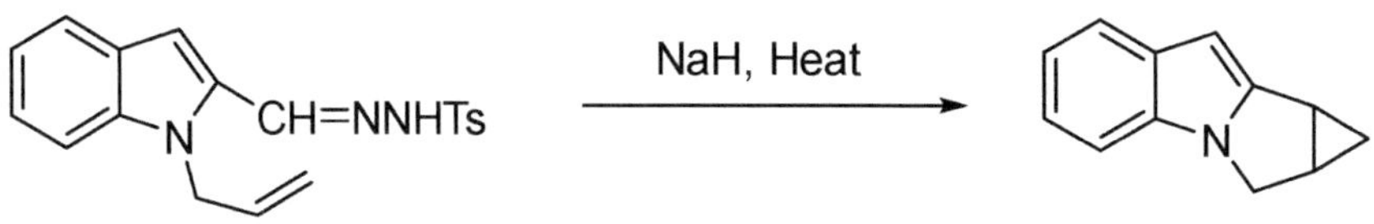

Figure 25.
Intramolecular carbene addition in the synthesis of tetracyclic indole.

6.2 Example 2

Cyclopentanes are prepared using intramolecular C-H insertions. The reactions involve the rhodium (II) catalyzed decomposition of α-diazo-β-keto esters as shown in **Figure 26**.

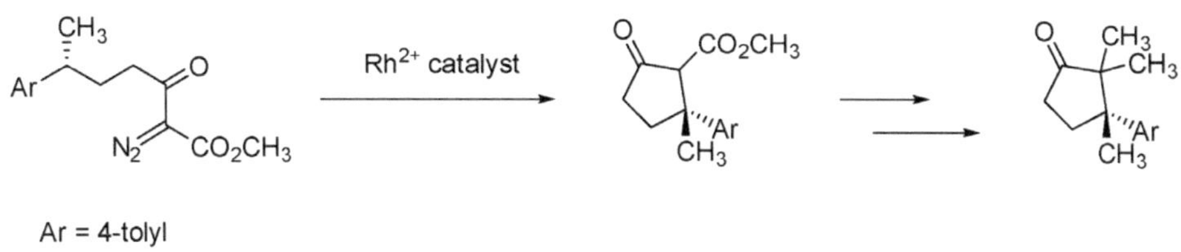

Figure 26.
Synthesis of cyclopentanones by intramolecular C-H insertion reaction.

6.3 Example 3

The β-lactam antibiotics are synthesized by using carbene reactions. This example as shown in **Figure 27** involves a Wolff rearrangement of a diazoketone to effect a ring contraction of five-membered ring.

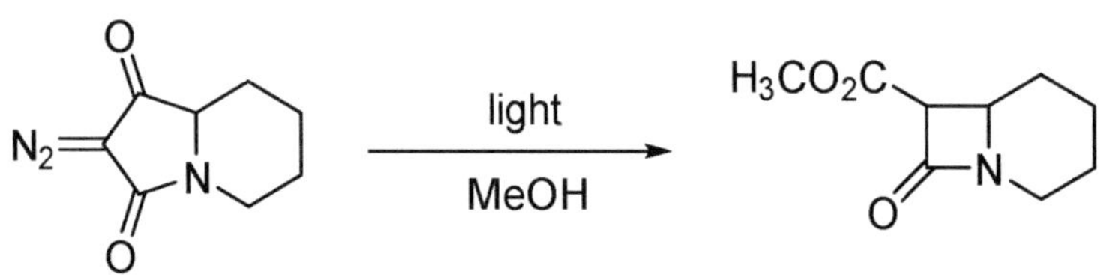

Figure 27.
Carbene reaction in the synthesis of β-lactam antibiotics.

6.4 Example 4

This example involves intramolecular O-H insertion reaction to synthesize heterocyclic compound (**Figure 28**).

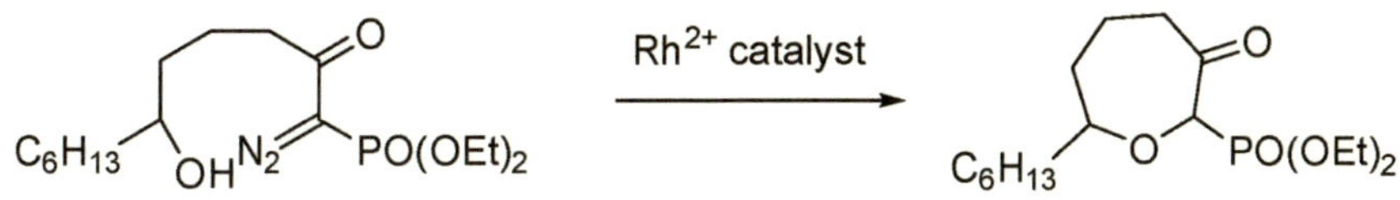

Figure 28.
Intramolecular O-H insertion.

6.5 Example 5: Arndt-Eistert reaction

Chain extension in carboxylic acid. Diazomethane is one of the reagents used to extend the chain in carboxylic acids. Carboxylic acids are converted to acid chlorides and treated with diazomethane without isolation to form α-diazoketones. The formed diazoketone easily turns into ketenes when heated. Finally, ketene reacts with water and gives carboxylic acid (**Figure 29**).

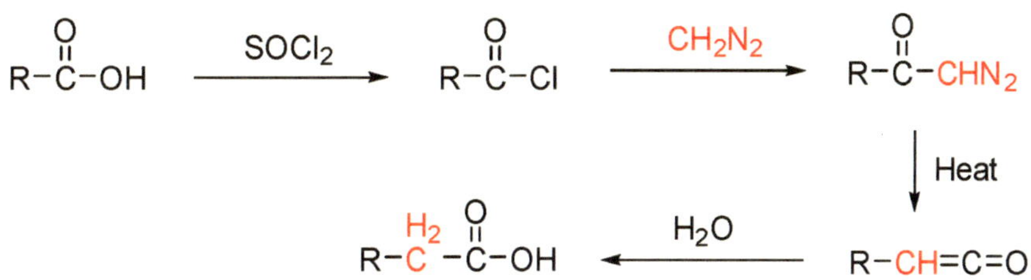

Figure 29.
Chain extension in carboxylic acid by using diazomethane.

Author details

Nuriye Tuna Subasi
Department of Food Engineering, Ahi Evran University, Kirşehir, Turkey

*Address all correspondence to: tunasubasi@gmail.com

References

[1] Moody CJ, Whitham GH. Reactive intermediates. 3rd ed. Oxford University Press; 1997. 91 p. ISBN-13: 978-0198556732

[2] Dumas JB, Peligot E. Mémoire sur l'esprit de bois et sur les divers composés éthérés qui en proviennent. Ann. Chim. Phys. 1835, 58, 5.

[3] Buchner E, Curtius T. Ueber die Einwirkung von Diazoessigäther auf aromatische Kohlenwasserstoffe. Ber. Dtsch. Chem. Ges. 1885, 8, 2377.

[4] Staudinger H, Kupfer O. Über reaktionen des methylens. iii. Diazomethan. Ber. Dtsch. Chem. Ges. 1912, 45, 501.

[5] Moss RA, Platz MS, Jones M Jr. Eds. Reactive Intermediate Chemistry. Wiley-Interscience: Hoboken, NJ, 2004.

[6] Doering WE, Hoffmann AK. The addition of dichlorocarbene to olefins. J Am. Chem. Soc. 1954, 76, 6162-6165.

[7] Balci M. Reaksiyon mekanizmaları. 3rd ed. Türkiye Bilimler Akademisi; 2012. 597 p.

[8] Kirmse W. Carbene Chemistry. Academic Press: New York, 1971.

[9] Moss RA, Platz MS, Jones, Jr. M. Reactive Intermediate Chemistry. Wiley-Interscience 2004. 1067 p. ISBN: 0-471-23324-2

[10] Jones M, Moss RA. Eds. Carbenes. Wiley Interscience: New York; 1973 (Vol. 1) and 1975 (Vol. 2)

[11] Brinker UH. Ed. Advanced in Carbene Chemistry. JAI Press: Greenwich; 1994, Vol. 1; JAI Press: Stamford; 1998, Vol. 2; Elsevier: Amsterdam; 2001, Vol. 3.

[12] Nemirowski A, Schreiner PR. Electronic Stabilization of Ground State Triplet Carbenes. J. Org. Chem. 2007;72 (25): 9533–9540. doi:10.1021/jo701615x. PMID 17994760.

[13] (a) Bourissou D, Guerret O, Gabbai FP, Bertrand G. Stable carbenes. Chem. Rev. 2000; 100: 39-91. (b) Müller PH, Rondan NG, Houk KN, Harrison JF, Hooper D, Willen BH, Liebman JF. Carbene singlet-triplet gaps. Linear correlations with substituent π-donation. J. Am. Chem. Soc. 1981; 103: 5049-5052.

[14] Sulzbach HM, Bolton E, Lenoir D, Schleyer PvR, Schaefer HF. Tetra-tert-butylethylene: An Elusive Molecule with a Highly Twisted Double Bond. Can It Be Made by Carbene Dimerization? J. Am. Chem. Soc. 1996; 118: 9908-9914

[15] (a) Feller D, Borden WT, Davidson ER. Dependence of the singlet-triplet splitting in heterosubstituted carbenes on the heteroatom electronegativity and conformation. Chem. Phys. Lett. 1980; 71: 22-26. (b) Bauschlicher CW, Schaefer HFIII, Bagus PS. Structure and energetics of simple carbenes methylene, fluoromethylene, chloromethylene, bromomethylene, difluoromethylene, and dichloromethylene. J. Am. Chem. Soc. 1977; 99: 7106-7110. (c) Baird NC, Taylor KF. Multiplicity of the ground state and magnitude of the T_1-S_0 gap in substituted carbenes. J. Am. Chem. Soc. 1978; 100: 1333-1338.

[16] Igau A, Grützmacher H, Baceiredo A, Bertrand G. Analogous α-α'-bis-carbenoid, triply bonded species: synthesis of a stable λ^3-phosphino carbene-λ^5-phosphaacetylene. J. Am. Chem. Soc. 1988; 110: 6463-6466.

[17] Wertz JE, Bolton JR. Electron Spin Resonance. McGraw-Hill: New York; 1972.

[18] Dougherty DA. Kinetics and Spectroscopy of Carbenes and Biradicals. Platz MS. Ed. Plenum Press: New York; 1990. 117 p.

[19] Trozzolo AM, Wasserman E. In Carbenes; Moss RA, Jones MJr. Eds. Wiley: New York; 1975; Vol. 2, p 185.

[20] Sander W, Bucher G, Wierlacher S. Carbenes in matrixes: spectroscopy, structure, and reactivity Chem. Rev. 1993; 93: 1583-1621.

[21] Igau A, Grützmacher H, Baceiredo A, Bertrand G. Analogous α-α'-bis-carbenoid, triply bonded species: synthesis of a stable λ^3-phosphino carbene-λ^5-phosphaacetylene. J. Am. Chem. Soc. 1988; 110: 6463-6466.

[22] Igau A, Baceiredo A, Trinquier G, Bertrand, G. [Bis(diisopropylamino) phosphino]trimethylsilylcarbene: A Stable Nucleophilic Carbene. Angew. Chem. Int. Ed. Engl. 1989; 28: 621-622.

[23] Arduengo AJ III, Harlow RL, Kline M. A stable crystalline carbene. J. Am. Chem. Soc. 1991; 113: 361-363.

[24] Curtius T. Ueber die Einwirkung von salpetriger Säure auf salzsauren Glycocolläther. Berichte der Deutschen Chemischen Gesellschaft. 1883; 16 (2): 2230–2231.

[25] Herzberg G, Shoosmith J. Spectrum and Structure of the Free Methylene Radical. Nature (London). 1959; 183: 1801-1802.

[26] Fulton JR, Aggarwal VK, de Vicente J. The use of tosylhydrazone salts as a safe alternative for handling diazo compounds and their applications in organic synthesis. Eur. J. Org. Chem. 2005;1479-1492.

[27] Bamford WR, Stevens TS. The decomposition of toluene-p-sulphonylhydrazones by alkali. J. Chem. Soc. 1952; 4735-4740.

[28] Davies HW, Schwarz M. The Effects of Hydrogen Bonding on the Absorption Spectra of Some Substituted Benzaldehyde Tosylhydrazone Anions. J. Org. Chem. 1965; 30: 1242-1244.

[29] Kristinsson H, Griffin GW. Photochemistry of Phenyloxiranes. II. New Precursor for Phenylcarbene. Angew. Chem. 1966; 88, 7: 1579–1580.

[30] Kristinsson H. Photochemische bildung von phenyl-methylcarben. Tetrahedron Letters 1966; 7 (21): 2343-2345.

[31] Petrellis PC, Dietrich H, Meyer E, Griffin GW. Photofragmentation of Epoxides. IV. Precursors for Phenylcyanocarbene. J. Am. Chem. Soc. 1967; 89 (8): 1967–1969.

[32] Petrellis PC, Griffin GW. Photofragmentation of oxirans. Precursors for phenylcyanocarbene and phenylmethoxycarbonylcarbene. Chem. Commun. 1967; 691-692.

[33] Temnikova TI, Stepanov IP, Semenova LO. Reactions of epoxy-compounds by a radical mechanism. Zh. Org. Khim. 1967; 3: 1708.

[34] Schmitz E. Dreiringe mit Zwei Heteroatomen. Springer Verlag, Berlin, Heidelberg, New York 1967. pp. 114-167.

[35] Milligan DE, Mann DE, Jacox ME, Mitsch RA. Infrared Spectrum of CF_2. J. Chem. Phy. 1964; 41: 1199-1203.

[36] Mitsch RA. Difluorodiazirine. III. Synthesis of Difluorocyclopropanes. J. Am. Chem. Soc. 1965; 87: 758-761.

[37] Moss RA. The photolysis of phenylbromodiazirine and the generation of phenylbromocarbene Tetrahedron Letters 1967; 8 (49): 4905-4909.

[38] Franzen V, Joschek HI. Untersuchungen über Carbene, VI.

Thermische Spaltung von Tetra-α-naphthyläthylen in Di-α-naphthylmethylen. Justus Liebigs Ann. Chem. 1960; 633: 7-13.

[39] Geuther A. Ann. 1862; 123: 121.

[40] Hine J, Dowell AM. Carbon Dihalides as Intermediates in the Basic Hydrolysis of Haloforms. III. Combination of Carbon Dichloride with Halide Ions. J. Am. Chem. Soc. 1954; 76: 2688-2692.

[41] Hine J, Langford PB. Methylene Derivatives as Intermediates in Polar Reactions. IX. The Concerted Mechanism for α-Eliminations of Haloforms. J. Am. Chem. Soc. 1957; 79: 5497-5500.

[42] Hine J. Divalent Carbon. Ronald Press Co., New York 1964.

[43] Makosza M, Wawrzyniewicz M. Reactions of organic anions. XXIV. Catalytic method for preparation of dichlorocyclopropane derivatives in aqueous medium. Tetrahedron Lett. 1969; 10 (53): 4659-4662.

[44] (a) Simmons HE, Smith RD. A new synthesis of cyclopropanes from olefins. J. Am. Chem. Soc. 1958; 80 (19): 5323–5324. (b) Simmons HE, Smith RD. A New Synthesis of Cyclopropanes. J. Am. Chem. Soc. 1959; 81 (16): 4256–4264.

[45] LeGoff E. Cyclopropanes from an Easily Prepared, Highly Active Zinc-Copper Couple, Dibromomethane, and Olefins. J. Org. Chem. 1964; 29 (7): 2048–2050.

[46] Furukawa J, Kawabata N, Nishimura J. A novel route to cyclopropanes from olefins. Tetrahedron Lett. 1966; 7 (28): 3353-3354.

[47] Skell PS, Woodworth RC. Structure Of Carbene, CH_2. J. Am. Chem. Soc. 1956; 78 (17): 4496-4497.

[48] Kirmse W. 100 Years of the Wolff Rearrangement. Eur. J. Org. Chem. 2002: 2193-2256.

Chapter 5

Late Transition Metal (LTM)-NHC Catalyzed Transformations of Renewable Chemicals to Fine Chemicals, Fuels, and Intermediates

Kurra Mohan, Bollikolla Hari Babu,
Khandapu Bala Murali Krishna, Kotra Vijay and Varala Ravi

Abstract

This title of the book chapter deals with the late transition metal-NHC (N-heterocyclic carbene) catalyzed transformations of renewable chemicals, i.e., bio-mass resources (carbohydrates/vegetable oils/natural products) into useful chemicals *via* oxidation, hydrogenation, dehydration, polymerization, hydrolysis, etc. along with brief introductory notes on late transition metals, carbenes, and renewable chemicals for better understanding to the reader.

Keywords: late transition metals, NHC, renewable chemicals, fine chemicals, fuels, intermediates

1. Introduction

Organocatalysis plays a pivotal role in the field of synthetic organic chemistry as well as the pharmaceutical industry through diversifying activation strategies owing to meeting the principles of green chemistry [1–4] in terms of cost-effectiveness, longevity, and less toxic compared to conventional transition metal catalysis [5–8]. In this regard, N-heterocyclic carbene (NHC) plays a major role in diversified organic transformations [9–11].

1.1 Renewable chemicals

Renewable chemicals or "bio-based chemicals" are chemicals obtained from renewable sources, such as agricultural feedstock, agricultural waste, organic waste products, biomass, and microorganisms [12]. In general, in chemical industries, processes include the utilization of fossil resources. As the need for energy consumption and population increasing, limited availability of fossil resources has become a risky task in the low or underdeveloped nations to perform trade. Henceforth, alternative renewable resources such as lignin, hemicellulose, cellulose, starch, and protein have become more focus of utility.

1.2 Carbene

The term "Carbene" refers to the presence of neutral bivalent carbon with six valence electrons in N-heterocyclic compounds (**Figure 1**). The first reported carbene (I) was by Bartrand et al. in 1988 [13], as resonance stabilized ylide form. After a few years, the first stable NHC was reported by Arduengo et al. as an imidazolium ring [14]. In NHC, the singlet state of carbene is more thermodynamically favorable than triplet carbene. Because nitrogen is present near to carbon of carbene, it lowers the energy of the highest occupied molecular orbital (HOMO) while it increases the energy of the lowest unoccupied molecular orbital's. The nucleophilicity of carbene also increases (**A**) not only above energy character but also presence of inductive effect, mesomeric and lone pair to vacant p-orbital favors singlet carbene. Most NHCs are based on imidazolium, triazolium, or thiazolium ring-containing molecules. NHCs dimerize reversible in the form of the Wanzlick equilibrium (**B**) [15–16].

Since the discovery of metal carbenes in 1964 by Fisher et al. [17], fascinating applications in both catalysis and synthesis are being observed [18].

1.3 Late transition metals

Late transition metals are on the right side of the d-block, from group 8 to 11 (and 12 if it is counted as transition metals) as shown in **Figure 2**.

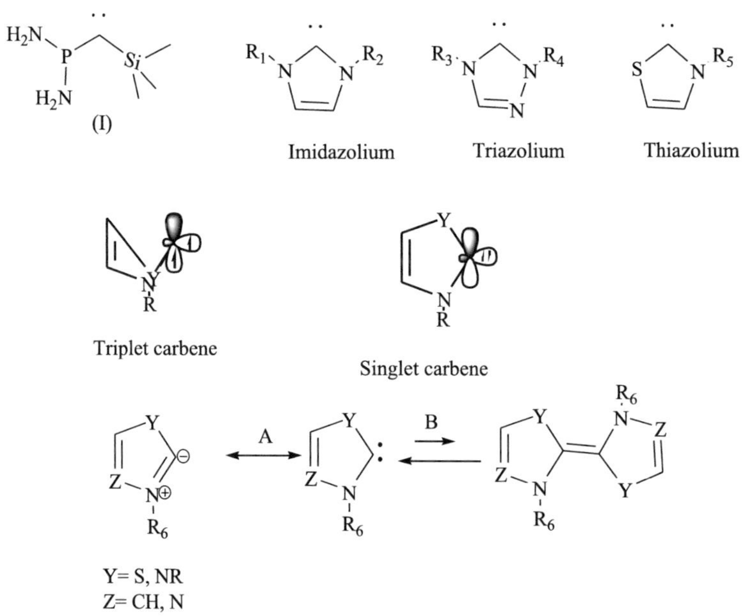

Figure 1.
The structure and stabilization of the first persistent carbene and NHC's.

Rb		Y	$5s^24d^2$ Zr $4d^4$	$5s^24d^3$ Nb $4d^5$	$5s^24d^4$ Mo $4d^6$	$5s^24d^5$ Tc $4d^7$	$5s^24d^6$ Ru $4d^8$	$5s^24d^7$ Rh $4d^9$	$5s^24d^{10}$ Pd $4d^{10}$	Ag	Cd	In		Xe
Cs		L a	$6s^25d^2$ Hf $5d^4$	$6s^25d^3$ Ta $5d^5$	$6s^25d^4$ W $5d^6$	$6s^25d^5$ Re $5d^7$	$6s^25d^6$ Os $5d^8$	$6s^25d^7$ Ir $5d^9$	$6s^25d^8$ Pt $5d^{10}$	Au	Hg	Tl		Rn

EARLY LATE

Figure 2.
Late transition metals.

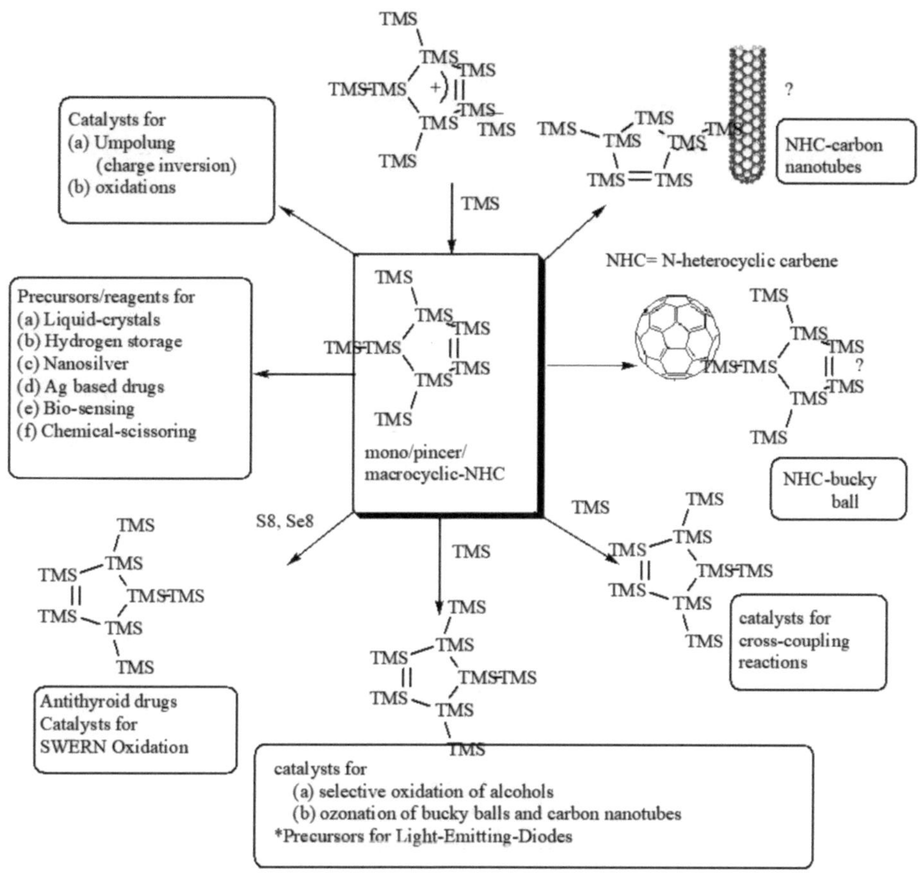

Figure 3.
Transmetalation route for the synthesis of carbenes.

1.4 Free carbine route

The general synthesis of carbene complexes involves the utilization of strong bases and harsh reaction conditions which involves high cost and more time.

1.5 Transmetalation route

It involves the transfer of the carbine fragment from a suitable metal center [generally Ag(I) or Cu(I)] to a precursor of the metal center of interest [19–22] as shown in **Figure 3**.

Even though, transmetalation method has operational simplicity but lacks atom economy. Hence, it is applied, in general, in scalable industrial processes.

2. Applications of late transition metal NHC's

2.1 CO_2 as building blocks

The exploitation of carbon dioxide as a renewable green source of carbon in organic synthesis is of continued interest. In this regard, late transition metal NHCs play a major role for the specified purpose.

2.1.1 Formylation of amines

The use of CO_2 for procuring C_1-containing molecules is an evolved methodology exploiting N-heterocyclic carbenes (NHCs) as efficient catalysts [23, 24]. NHCs promoted the formylation of a wide scope of N-H bonds, with CO_2 and hydrosilanes (**Figure 4**) [25].

2.1.2 Carboxylation of terminal alkynes

Yu and Zhang [26] developed a Cu-NHC catalyzed conversion of CO_2 to carboxylic acids in good to excellent yields under ambient conditions with wide substrate/ functional group tolerance (**Figure 5**).

2.1.3 Methylation of amines

Olivier et al. have designed using CO_2 as a $C1^-$ building block for the catalytic methylation of amines using simple zinc salts and ligands (**Figure 6**) [27].

2.1.4 Insertion of CO_2 into terminal alkynes via copper bis-NHC

Silver bis-NHC has exhibited better performance than Copper bis-NHC towards the carboxylation of terminal alkynes using Cs_2CO_3 as an additive (**Figure 7**) [28].

2.1.5 Carboxylative cyclization of propargylamine

Tahani et al. synthesized dinuclear gold (I) complexes and investigated the carboxylative cyclization of propargylamine (PPA) (**Figure 8**) [29].

2.2 Oxidation

2.2.1 Dehydrogenative oxidation of alcohols

Ir-NHC complexes were synthesized in aqueous media for the oxidation of secondary alcohols to ketones. In addition, primary alcohols were transformed to carboxylic acids in the absence of a base [30].

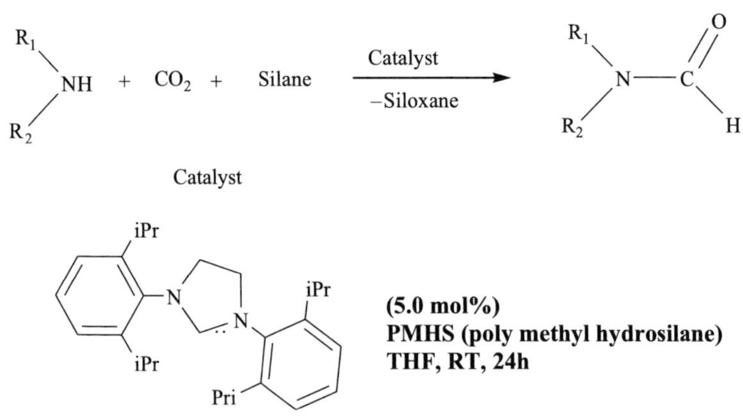

Figure 4.
Formylation of amines with CO_2 and hydrosilanes.

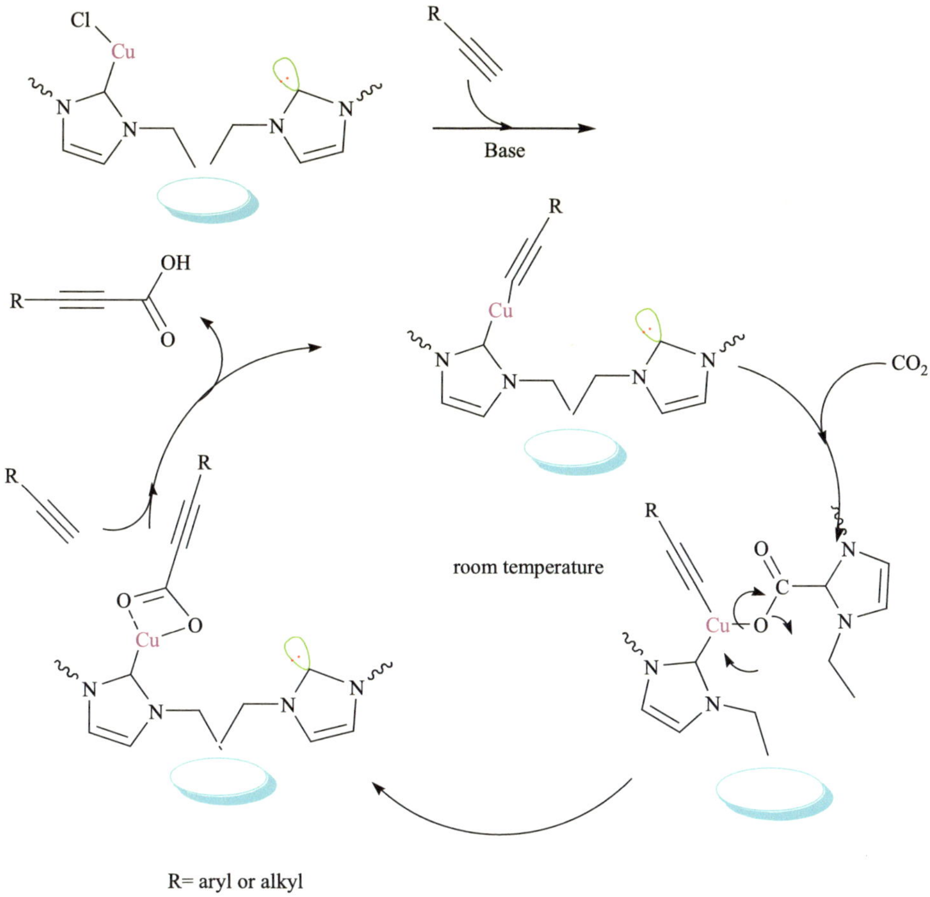

Figure 5.
Mechanistic approach for carboxylation of terminal alkynes.

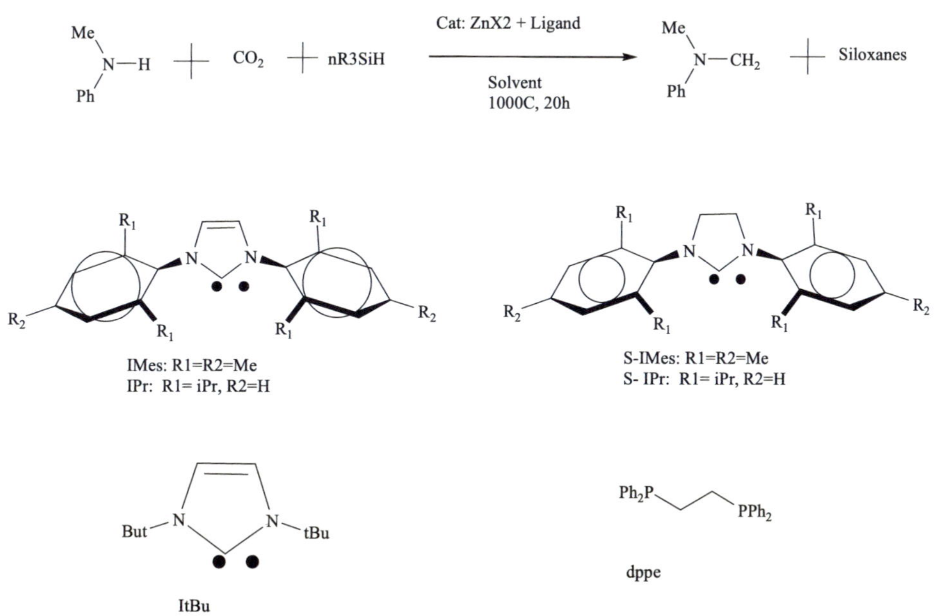

Figure 6.
N-methylation of amines.

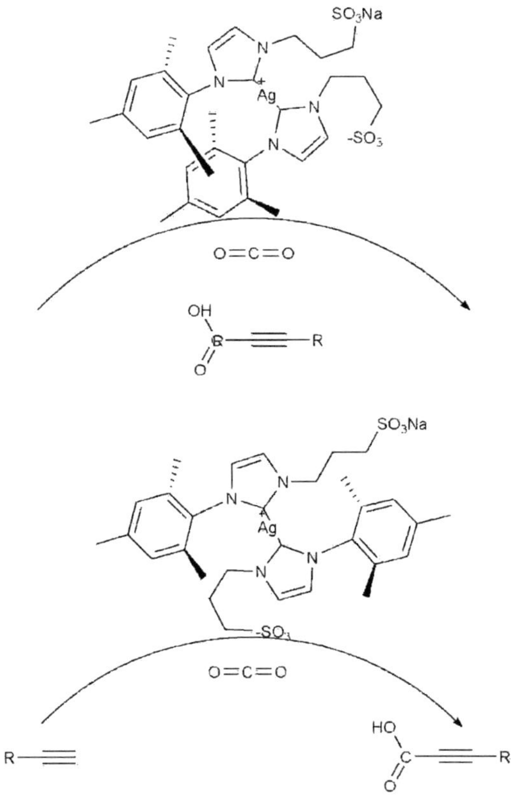

Figure 7.
Carboxylation of terminal alkynes.

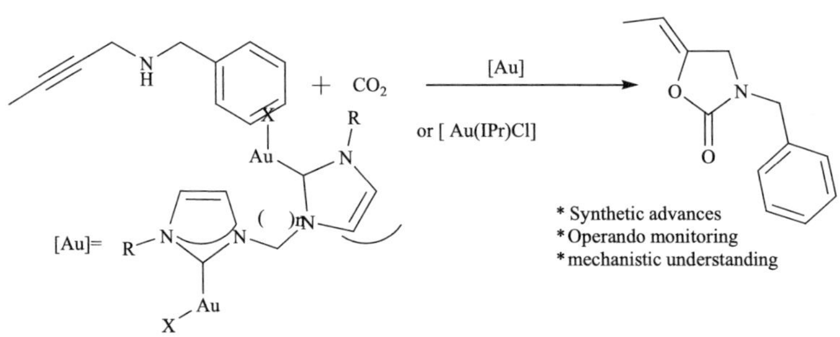

Figure 8.
Carboxylative cyclization of propargylamine.

2.2.2 Oxidation of bio-polyols to lactic acid

Lactic acid has prominent applications in bio-plastics manufacturing. A recyclable NHC-iridium coordination polymer with a porous structure can oxidize a wide range of bio-polyols such as sorbitol to prepare lactic acid with superior selectivity and reactivity [31].

2.2.3 Dehydrogenative catalysis using alcohols

Huang et al. reported LTM-NHCs for the conversion of alcohols into aldehydes or ketones through acceptors alcohol dehydrogenation (AAD). In addition, they successfully demonstrated oxidative coupling of alcohols to form C-O, C-C, and C-N/C=N bond formations (**Figure 9**) [32].

2.2.4 Dehydrogenation of sugar alcohols

Manas and Campos et al. [33] reported Ir-NHC catalyzed oxidative protocol for the selective conversion of sorbitol, xylitol, and other polyols into lactic acid (**Figure 10**).

2.3 Dehydration

2.3.1 Cp*IrCl2(NHC) in hydrogen transfer initiated dehydration (HTID)

A recyclable Cp*IrCl$_2$(NHC) (Cp* = pentamethylcyclopentadienyl) complex in ionic liquid could covert glycerol into 1,3-propanediol and subsequently to propionaldehyde by hydrogen transfer initiated dehydration (HTID) in excellent yields in the presence of air (**Figure 11**) [34, 35].

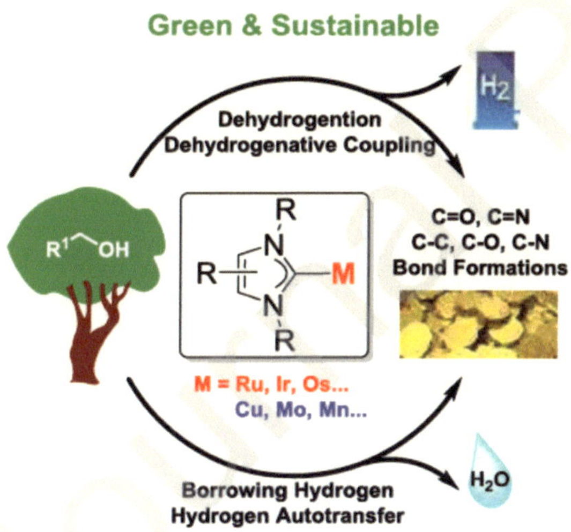

Figure 9.
Dehydrogenative catalysis using alcohols.

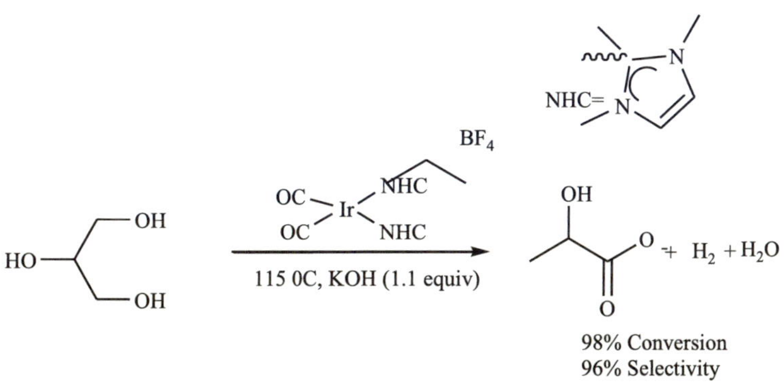

Figure 10.
Oxidation of sugar alcohols to lactic acid.

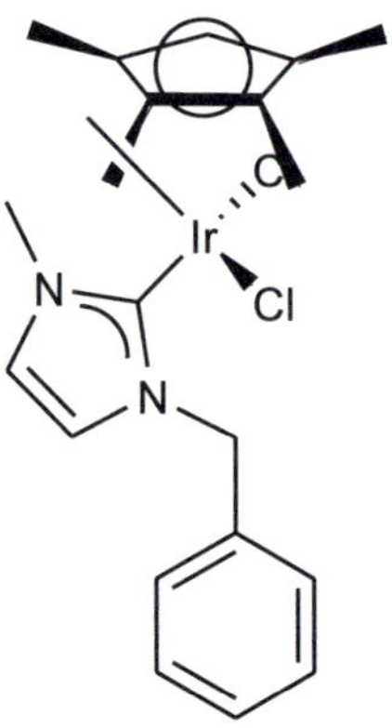

Figure 11.
*Cp*IrCl2(NHC) in hydrogen transfer initiated dehydration (HTID).*

2.3.2 Fructose to 5-hydroxymethylfurfural (HMF)

A new heterogeneous and recyclable Fe-NHCs immobilized on mesoporous expanded starch and Starbon™ 350 could be utilized successfully for the effective dehydration of fructose to HMF [36].

2.4 Reduction/hydrogenation

2.4.1 Hydrogenolysis of aryl ethers using Ni-NHC

Ni-NHC complex in the presence of a suitable base (NaO^tBu) could effectively convert C-O bonds in lignin to various useful scaffolds useful in biomass conversion [37]. Hartwig et al. mechanically investigated the reduction of diaryl ethers to corresponding phenols (**Figure 12**) [38].

2.4.2 Transfer hydrogenation using Ir-NHC

Using water soluble Ir-NHCs proved that glycerol can be exploited as a hydrogen donor to convert a biomass-derived phytochemical, levulinic acid, to selectively produce γ-hydroxyvaleric acid (GHV) and lactic acid (LA) [39].

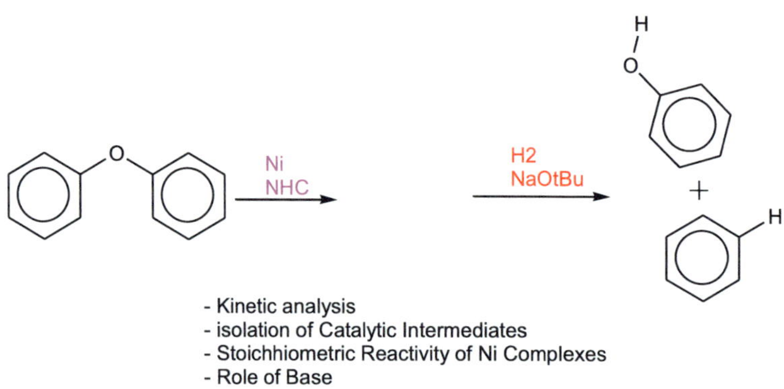

Figure 12.
Hydrogenolysis of diaryl ethers.

2.4.3 Iridium-based hydride transfer catalysts

Lu et al. reported homogeneous Ir-NHC catalysts, which can be utilized for the storage of H_2 and fine chemicals through hydride transfer catalysis [40] (**Figure 13**).

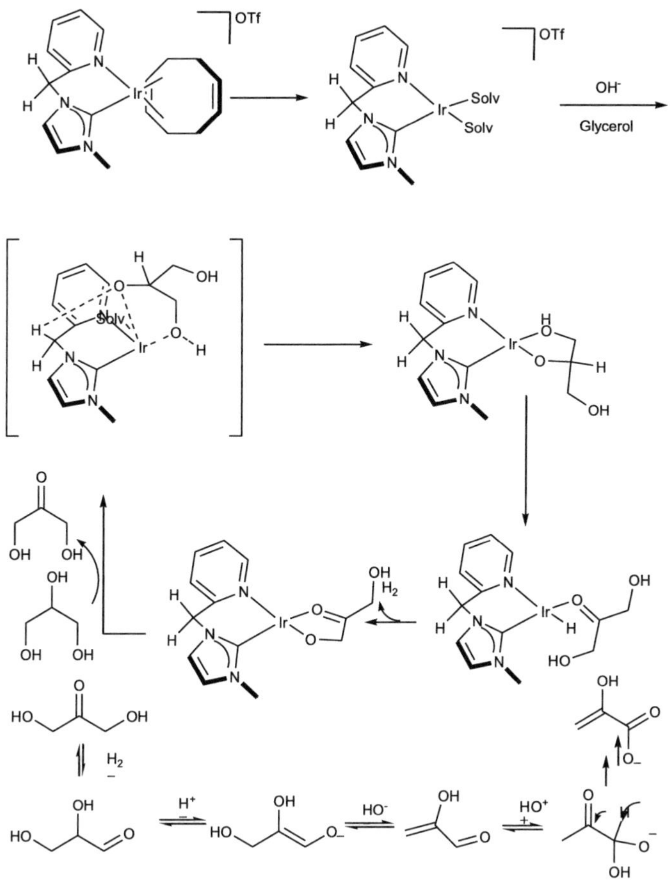

Figure 13.
Iridium-based hydride transfer catalysts.

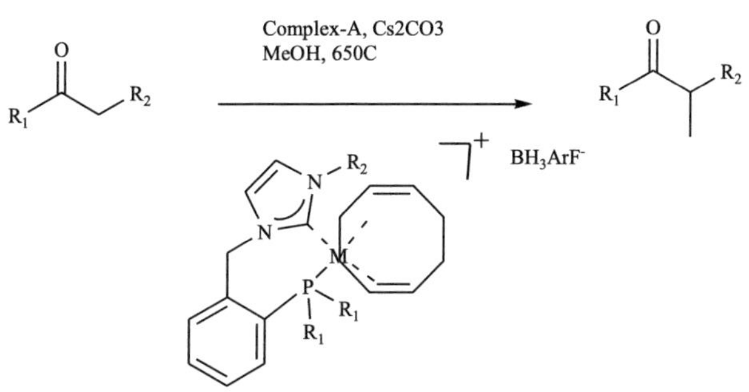

Figure 14.
Conversion of sugars into heterocycles.

2.5 Miscellaneous organic transformations

2.5.1 Sugars to heterocycles

Zhang and Yong developed a synthetic protocol employing Cr-NHC along with ionic liquid for the selective production of 5-hydroxymethylfurfural from glucose and fructose (**Figure 14**) [41].

3. Conclusion

In this book chapter, authors tried to emphasize the applications of "Late Transition Metal" (LTM)-NHC catalyzed organic transformations as given in a nutshell below:

Oxidation of carbohydrates: To develop carbohydrate oxidation products as a useful alternative to those derived from petrochemical sources.

Hydrogenation of carbohydrates/fatty acids: This objective concerns the development of LTM-NHC catalysts for the hydrogenation of carbohydrates and unsaturated vegetable oils.

Dehydration/hydrolysis of carbohydrates/fatty acids: Development of dehydration/hydrolysis of carbohydrates/fatty acids with LTM-NHC catalysts to obtain fine chemicals and fuel intermediates.

Polymerization with renewable resources: This objective deals with the application of LTM-NHC catalysts in the polymerization of natural monomers of renewable chemicals or monomers derived from renewable resources to synthetic polymers (polymerization of lactic acid, glucose, glycerol, terpenes, etc.).

The present research is directed towards the conversion of methanol to H_2 and CO_2 using LTM-NHC catalysis.

$$CH_3OH(g)+H_2O \xrightarrow{\text{LTM-NHC Catalyst}} 3H_2(g)+CO_2(g) \quad (1)$$

We do hope this compilation on very important LTM-NHC applications would help wide readers among synthetic organic chemists.

Acknowledgements

Dr. RV is thankful to Dr. Ch.V. Rajasekhar, Scrips Pharma, and Dr. P.G. Kiran, Swastha BioSciences for their continued support.

Conflict of interest

The authors declare no conflict of interest.

Author details

Kurra Mohan[1], Bollikolla Hari Babu[2], Khandapu Bala Murali Krishna[2], Kotra Vijay[3] and Varala Ravi[4*]

1 Department of Pharmaceutical Chemistry, Telangana University, Telangana, India

2 Department of Chemistry, Acharya Nagarjuna University, Guntur, AP, India

3 Faculty of Pharmacy, Quest International University, Ipoh, Perak, Malaysia

4 Scrips Pharma, Hyderabad, Telangana, India

*Address all correspondence to: ravivarala@gmail.com

References

[1] Dalko PI. Enantioselective Organocatalysis. KGaA, Weinheim: WILEY-VCH Verlag Gmbh & Co.; 2007

[2] Berkessel A, Groger H. Asymmetric Organocatalysis—From Biomimetic Concepts to Applications in Asymmetric Synthesis. KGaA, Wienheim: WILEY-VCH Verlag GmbH & Co.; 2006

[3] Aleman J, Cabrera S. Applications of asymmetric organocatalysis in medicinal chemistry. Chemical Society Reviews. 2013;**42**:774-793. DOI: 10.1039/C2CS35380F

[4] Busacca CA, Fandrick DR, Song JJ, Senanayake CH. The growing impact of catalysis in the pharmaceutical industry. Advanced Synthesis and Catalysis. 2011;**353**:1825-1864. DOI: 10.1002/adsc.201100488

[5] Parmar D, Sugiono E, Raja S, Rueping M. Complete field guide to asymmetric BINOL-phosphate derived brønsted acid and metal catalysis: History and classification by mode of activation; Brønsted acidity, hydrogen bonding, ion pairing, and metal phosphates. Chemical Reviews. 2014;**114**:9047-9153. DOI: 10.1021/cr5001496

[6] Ishikawa T. Superbases for Organic Synthesis: Guanidines, Amidines, Phosphazenes and Related Organocatalysts. New York: Wiley; 2009

[7] Zhang Z, Schreiner PR. (Thio)urea organocatalysis—What can be learnt from anion recognition? Chemical Society Reviews. 2009;**38**:1187-1198. DOI: 10.1039/B801793J

[8] Hashimoto T, Maruoka K. Recent development and application of chiral phase-transfer catalysts. Chemical Reviews. 2007;**107**:5656-5682. DOI: 10.1021/cr068368n

[9] Federsel H-J. Asymmetry on large scale: The roadmap to stereoselective processes. Nature Reviews. Drug Discovery. 2005;**4**:685-697. DOI: 10.1038/nrd1798

[10] Breuer M, Ditrich K, Habicher T, Hauer B, Keseler M, Sturmer R, et al. Industrial methods for the production of optically active intermediates. Angewandte Chemie International Edition. 2004;**43**:788-824. DOI: 10.1002/anie.200300599

[11] Fiorani G, Guo W, Kleij AW. Sustainable conversion of carbon dioxide: The advent of organocatalysis. Green Chemistry. 2015;**17**:1375-1389. DOI: 10.1039/C4GC01959H

[12] Chandel AK, Garlapati VK, Kumar SPJ, Hans M, Singh AK, Kumar S. The role of renewable chemicals and biofuels in building a bioeconomy. Biofuels, Bioproducts and Biorefining. 2020;**14**(4):830-844. DOI: 10.1002/bbb.2104

[13] Igau A, Grutzmacher H, Baceiredo A, Bertrand G. Analogous alpha, alpha.'-bis-carbenoid, triply bonded species: Synthesis of a stable lambda.3-phosphino carbene-. lambda.5-phosphaacetylene. Journal of the American Chemical Society. 1988;**110**:6463-6466. DOI: 10.1021/ja00227a028

[14] Arduengo AJ, Harlow RL, Kline M. A stable crystalline carbene. Journal of the American Chemical Society. 1991;**113**:361-363. DOI: 10.1021/ja00001a054

[15] Herrmann WA, Kocher C. N-heterocyclic carbenes. Angewandte Chemie International Edition. 1997;**36**:2162-2187. DOI: 10.1002/anie.199721621

[16] Hopkinson MN, Richter C, Schedler M, Glorius F. An overview of N-heterocyclic carbenes. Nature.

2014;**510**:485-496. DOI: 10.1038/nature13384

[17] Fischer EO, Maasbol A. On the existence of a tungsten carbonyl carbene complex. Angewandte Chemie. 1964;**76**:645. DOI: 10.1002/anie.196405801

[18] Fischer EO, Maasbol A. Übergangsmetall-Carben-Komplexe, II. Phenylmethoxycarben- und Methylmethoxycarben-pentacarbonyl-chrom, -molybdän, -wolfram und -cyclopentadienyl-dicarbonyl-mangan. Chemistry. 1967;**100**(2445). DOI: 10.1002/cber.19671000744

[19] Wang HMJ, Lin IJB. Facile synthesis of silver (I)-carbene complexes. Useful carbene transfer agents. Organometallics. 1998;**17**:972-975. DOI: 10.1021/om9709704

[20] Furst MRL, Cazin CSJ. Copper N-heterocyclic carbene (NHC) complexes as carbene transfer reagents. Chemical Communications. 2010;**46**:6924-6925. DOI: 10.1039/C0CC02308F

[21] Díez-González S. N-Heterocyclic Carbenes. From Laboratory Curiosities to Efficient Synthetic Tools. 2nd ed. London: Royal Society of Chemistry; 2017

[22] Scattolin T, Nolan SP. Synthetic routes to late transition metal—NHC complexes. Trends in Chemistry. 2020;**2**(8). DOI: 10.1016/j.trechm.2020.06.001

[23] Anis T, Blondiaux E, Xavier F, Thibault C. Reductive functionalization of CO_2 with amines: An entry to formamide, formamidine and methylamine derivatives. Green Chemistry. 2015;**17**:157-168. DOI: 10.1039/C4GC01614A

[24] Riduan SN, Zhang Y, Ying JY. Conversion of carbon dioxide into methanol with silanes over N-heterocyclic carbene catalysts. Angewandte Chemie International Edition. 2009;**48**:3322. DOI: 10.1002/anie.200806058

[25] Jacquet O, Gomes CDN, Ephritikhine M, Cantat T. Recycling of carbon and silicon wastes: Room temperature formylation of N-H bonds using carbon dioxide and polymethylhydrosiloxane. Journal of the American Chemical Society. 2012;**134**:2934. DOI: 10.1021/ja211527q

[26] Yu D, Zhang Y. Copper, and copper-N-heterocyclic carbene-catalyzed C-H activating carboxylation of terminal alkynes with CO_2 at ambient conditions. Proceedings of the National Academy of Sciences of the United States of America. 2010;**107**(47):20184-20189. DOI: 10.1073/pnas.1010962107

[27] Jacquet O, Xavier F, Christophe DNG, Thibault C. CO_2 as a $C1^-$ building block for the catalytic methylation of amines. Chemical Science. 2013;**4**:2127. DOI: 10.1039/c3sc22240c

[28] Velázquez HD, Wu Z-X, Vandichel M, Verpoort F. Inserting CO_2 into terminal alkynes via bis-(NHC)-metal complexes. Catalysis Letters. 2017;**47**(2):463-471. DOI: 10.1007/s10562-016-1920-5

[29] Bayrakdar TACA, Nahra F, Davis JV, Mohan MG, Captain B, Temprado M, et al. Dinuclear Gold(I) complexes bearing alkyl-bridged bis(Nheterocyclic carbene) ligands as catalysts for carboxylative cyclization of propargylamine: Synthesis, structure, and kinetic and mechanistic comparison to the mononuclear complex [Au(IPr)Cl]. Organometallics. 2020;**39**(15):2907-2916. DOI: 10.1021/acs.organomet.0c00404

[30] Fujita K, Tamura R, Yuhi T, Yoshida M, Onoda M, Yamaguchi R.

Dehydrogenative oxidation of alcohols in aqueous media catalyzed by a water-soluble dicationic iridium complex bearing a functional N-heterocyclic carbene ligand without using base. ACS Catalysis. 2018;**7**(10):7226-7230. DOI: 10.1021/acscatal.7b02560

[31] Wu J, Shen L, Duan S, Chen Z-N, Zheng Q, Liu Y, et al. Selective Catalytic dehydrogenative oxidation of bio-polyols to lactic acid. Angewandte Chemie International Edition. 2020;**59**(33):13871-13878. DOI: 10.1002/anie.202004174

[32] Huang M, Liu J, Li Y, Lan X-B, Su P, Zhao C, et al. Recent advances on N-heterocyclic carbene transition metal complexes for dehydrogenative catalysis using alcohols. Catalysis Today. 2020;**370**:114-141. DOI: 10.1016/j.cattod.2020.10.022

[33] Manas MG, Campos J, Sharninghausen LS, Lin E, Crabtree RH. Selective catalytic oxidation of sugar alcohols to lactic acid. Green Chemistry. 2015;**17**:594-600. DOI: 10.1039/C4GC01694G

[34] Wang Y-M, Lorenzini F, Rebros M, Saunders GC, Marr AC. Combining bio- and chemo-catalysis for the conversion of bio-renewable alcohols: Homogeneous iridium catalysed hydrogen transfer initiated dehydration of 1,3-propanediol to aldehydes. Green Chemistry. 2016;**18**(6):1751-1761. DOI: 10.1039/c5gc02157j

[35] Ma Y, Wang Y-M, Morgan PJ, Jackson RE, Liu X, Saunders GC, et al. Designing effective homogeneous catalysis for glycerol valorisation: Selective synthesis of a value-added aldehyde from 1,3-propanediol via hydrogen transfer catalysed by a highly recyclable, fluorinated Cp*Ir(NHC) catalyst. Catalysis Today. 2018;**307**:248-259. DOI: 10.1016/j.cattod.2017.09.036

[36] Matharu AS, Ahmed S, Almonthery B, Macquarrie DJ, Lee Y-S, Kim Y. Novel Starbon™/HACS-supported N-heterocyclic carbene-iron (III) catalyst for efficient conversion of fructose to HMF. ChemSusChem. 2017;**11**(4):716-725. DOI: 10.1002/cssc.201702207

[37] Xu L, Chung LW, Wu Y-D. Mechanism of Ni-NHC catalyzed hydrogenolysis of aryl ethers: Roles of the excess base. ACS Catalysis. 2016;**6**:483-493. DOI: 10.1021/acscatal.5b02089

[38] Saper NI, Hartwig JF. Mechanistic investigations of the hydrogenolysis of diaryl ethers catalyzed by nickel complexes of N-heterocyclic carbene ligands. Journal of the American Chemical Society. 2017;**139**:17667-17676. DOI: 10.1021/jacs.7b10537

[39] Wang K, Heltzel J, Evan S, Culley K, Gabriel L, Adelina V. Transfer hydrogenation of levulinic acid from glycerol and ethanol using water-soluble iridium N-heterocyclic carbene complexes. Journal of Organometallic Chemistry. 2020;**919**:121310. DOI: 10.1016/j.jorganchem.2020.121310

[40] Lu Z, Cherepakhin V, Demianets I, Lauridsen PJ, Williams TJ. Iridium-based hydride transfer catalysts: From hydrogen storage to fine chemicals. Chemical Communications. 2018;**54**:7711. DOI: 10.1039/c8cc03412e

[41] Yong G, Zhang Y, Ying JY. Efficient catalytic system for the selective production of 5-hydroxymethylfurfural from glucose and fructose. Angewandte Chemie International Edition. 2008;**47**:9345-9348. DOI: 10.1002/anie.200803207

Chapter 6

Novel N-Heterocyclic Carbene Silver (I) Complexes: Synthesis, Structural Characterization, Antimicrobial, Antioxidant and Cytotoxicity Potential Studies

Ichraf Slimani, Khaireddine Dridi, Ismail Özdemir, Nevin Gürbüz and Naceur Hamdi

Abstract

Nowadays, N-heterocyclic carbene-based silver-complexes Ag(I) have been widely used as an organometallic drug candidate in medicinal and pharmaceutical chemistry researches due to their low toxicity. Due to the success of Ag(I) complexes in biological applications, interest in the synthesis and applications of such compounds is increasing rapidly. Therefore, in this study, a series of unsymmetrical N,N-disubstituted benzimidazolium salts were synthesized as N-heterocyclic carbene (NHC) (2a-2j). The interaction of these benzimidazolium salts having their two nitrogen atoms substituted by bulky groups with Ag_2O in DMF has been carried out to afford Ag(I) complexes and characterized by ^{1}H NMR, ^{13}C NMR, FT-IR and elemental analyses. The antimicrobial activity of Ag(I) complexes was tested against some standard culture collections of Gram-negative, Gram-positive bacterial strains and Fungal strains, which are the most frequently isolated among the society and hospital-acquired infectious microorganisms as potential metallopharmaceutical agents. The Ag-NHC complexes showed effective antimicrobial activity against microorganisms with MIC values between 0.0024 and 1.25 mg/ml. Moreover, these Ag-NHC complexes exhibited significant antioxidant activities. In addition, of benzimidazoles salts 2,4 and Ag(I) complexes 3,5 were screened for their antitumor activity. The highest antitumor activity was observed for **3e** and **3d** Complexes.

Keywords: N-heterocyclic carbene, benzimidazolium salts, silver (I)-NHC complexes, antimicrobial, antioxidant and antitumor activities

1. Introduction

N-Heterocyclic carbenes (NHCs) are nitrogen-based heterocyclic compounds containing a divalent carbon atom. Previously, many researchers tried numerous synthetic methods to isolate the stable NHCs, but they were not successful until the first stable free-carbene was isolated in 1991 as a crystal solid by Arduengo and coworkers [1]. Since then, the number of studies in carbene chemistry has increased considerably, and has become stable in research laboratories throughout the world. Today, NHCs are one of the important classes of ligands for coordination chemistry. NHCs have strong σ-donating but, weak π-accepting properties, which show excellent support to stabilize various oxidation states of transition-metal. Also, they can provide steric and electronic properties for the optimal design of transition-metal complexes [2–8]. The modification at the nitrogen atoms of the NHCs significantly influence the reactivity and binding affinity of the ligand; thus, NHCs make the strong metal-carbon bond with different metals. Transition-metal complexes of NHCs are used as strong-, reactive- and selective-catalysts in many chemical reactions. Initially, the metal-NHC complexes were used extensively as a catalyst in organic transformations such as C-C, C-heteroatom cross-couplings, and C-H functionalization [9–12]. Also, in recent years, transition metal-NHC complexes containing Au, Pd, Cu, Ru, Pt, Ag, Rh metals have been widely used in medicine and pharmacy as the potential metallopharmaceutical agents against AMR [13–16]. Although, most of the organometallic drug research has focused on platinum- and gold-containing compounds, carbene-based silver-compounds stand out in the class of organometallic drugs owing to their low toxicity, easy synthesis, stability and limited possibility of side effects. Ag(I) complexes possess several properties, ranging from antibacterial, anticancer, anti-inflammatory and antiseptic to anti-neoplastic activity [17]. Ag(I) complexes have been recently at the focal point with increased attention due to their usually strong antimicrobial and anticancer properties, and have more effective than other transition-metal complexes, and also, they have low toxicity for humans. Ag(I) complexes also promise to be agents capable of overcoming AMR and beating antibiotic resistant bacteria, fungi and parasites [18–34]. Heterocyclic molecules are an important family of organic chemistry with a wide range of applications [35]. Although, this family is generally known for its agrochemical and veterinary applications, it is also used as a corrosion inhibitor, sanitizer, and dyestuff [36]. Many heterocyclic molecules such as favipiravir have also important pharmaceutical applications [37, 38]. For example, ribavirin is an N-heterocyclic molecule that is used in the treatment of COVID-19 [39]. The reasonable results obtained from bioactivity studies have enabled them to be a family that is frequently used in pharmaceutical chemistry [40]. NHCs, which are known for their high catalytic activity, are easily synthesized and modified [41, 42]. NHC metal complexes have become a popular research area with the frequent usage of metals in drug molecules. In our previous works, we concluded that the presence of electron-donating and bulky substituents attached to the nitrogen of the carbene ligand increases the antimicrobial activity of the silver complexes. These exciting results have led us to further investigate the antimicrobial properties of silver-NHCs. In this regard, herein, we now report the synthesis of novel NHC salts and their Ag(I) complexes and investigate their antimicrobial, antioxidant and cytotoxic activities. All salts and complex structures were characterized by elemental analysis, Fourier transforms infrared (FTIR), 1H and 13C nuclear magnetic resonance (NMR) spectroscopies.

2. Results and discussion

2.1 Preparation of benzimidazolium salts

Nitrogen-containing heterocyclic compounds received great attention because of their wide range of catalytic and pharmacological properties in organometallic chemistry. In this study, benzimidazoles salts (**2a-j**) prepared by the reaction of N-(isobutyl)-benzimidazole (1) with various aryl chloride in DMF at 80°C for 24 h. The reaction pathway is shown in **Figure 1**.

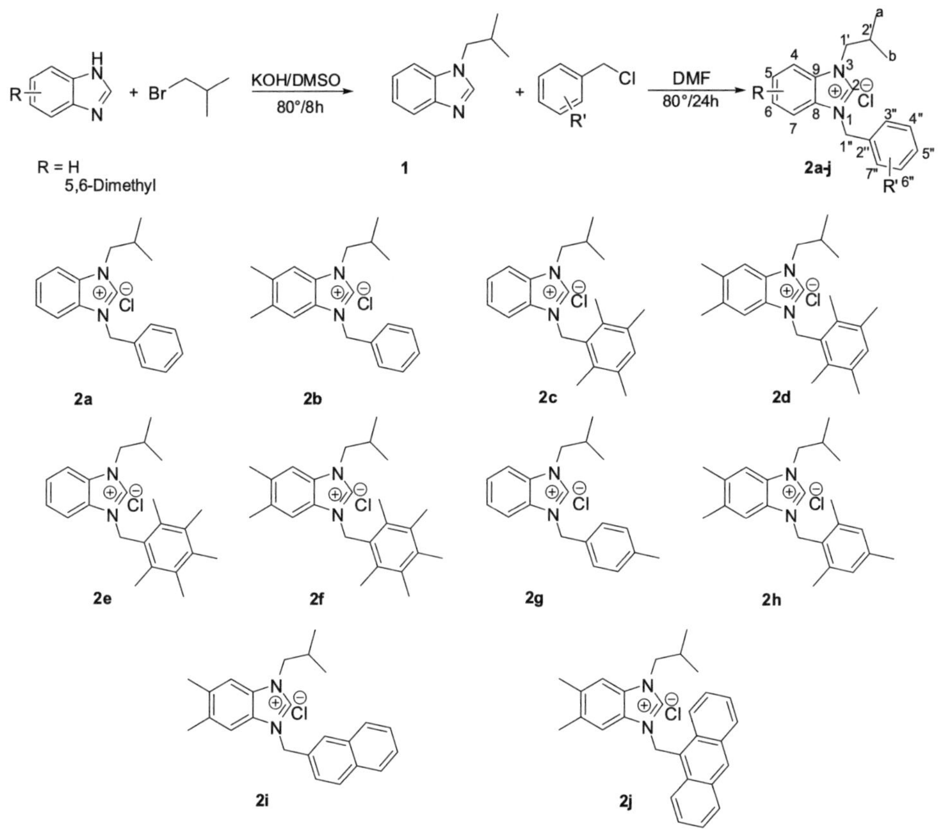

Figure 1.
Synthesis of the benzimidazoles salts (2a-j).

The NMR spectra of all compounds were run in δ-CDCl3. The acidic protons (NCHN) of the benzimidazolium salts (**2a-j**) were detected in the ^{1}H NMR spectra at 12.07, 11.81, 11.44, 11.08, 11.29, 10.48, 12.05, 11.34, 11.52, and 11.95 ppm, respectively, as a typical singlet. These are in agreement with values in the associated literature [43–49]. The methyl protons of the isopropyl group on the benzimidazolium salts (**2a-j**) were observed between 0.98 and 1.06 in the form of doublets, whereas the methyl protons of the benzimidazolium salts (**2a-j**) were signaled at 2.24–2.44 ppm as singlets. The isopropyl group $H_{2'}$ protons on all the benzimidazolium salts were visualized as septets in the interval 2.34–2.44 ppm. Also, in the ^{1}H NMR spectra of (**2a-j**), the $H_{1'}$ protons appeared between 4.32–4.51 ppm while the $H_{1''}$ protons were detected as typical singlet between 5.80–6.90 ppm. The signals detected in the range of 6.94–8.64 ppm are assigned to the aromatic protons of benzimidazolium salts (**2a-j**). In ^{13}C NMR spectra, the N-HCN (C_2) carbene peak of benzimidazole salts (**2a-j**) was assigned between 141.91–144.02 ppm.

Ag_2O and Benzimidazolium salts (**2a-j**) were reacted in dichloromethane at room temperature under dark and Ag(I)-NHC complex (**3a-j**) was obtained in very good yields. The Ag(I) complexes (**3a-j**) have good solubility in polar solvents and are stable in the air and towards the moisture. The synthetic route for the synthesis of Ag(I)-NHC complex (**3a-j**) is shown in **Figure 2**. In the ^{1}H NMR spectra, the acidic imino proton of benzimidazolium salts (NCHN) were not observed between δ 10.48–12.07 ppm. Similarly, in the ^{13}C NMR spectra, imino carbon of benzimidazolium salts (NCHN) was not observed between δ 141–144 ppm.

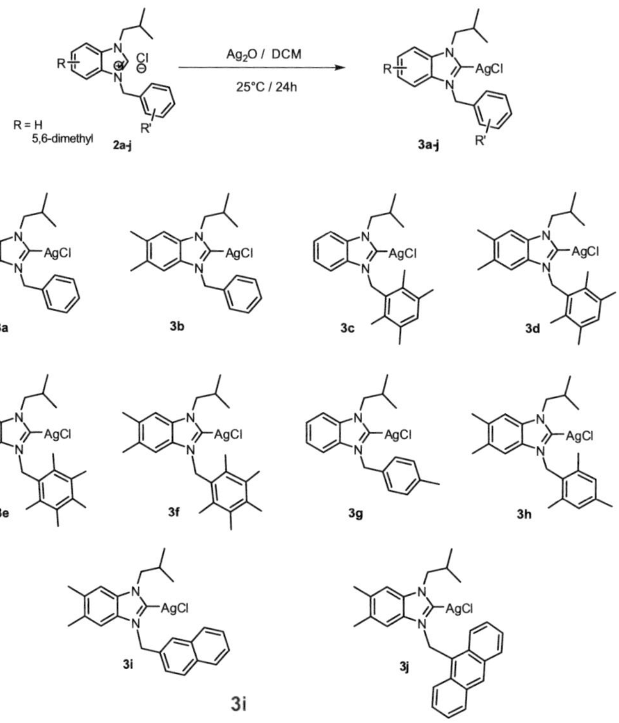

Figure 2.
Synthesis of silver(I) complexes 3a-j.

At the same time, the formation of the Ag(I) complexes (**3a-j**) was proven by IR spectra, which showed CN bond vibrations in the range of 1400–1591 cm^{-1}. The antibacterial and antioxidant activities of all the synthesized benzimidazolium salts (**2a-j**) and their corresponding Ag(I) complexes (**3a-j**) were evaluated as per details given in the following text.

3. Biological activities

It is known that the number of silver centers plays an important role in biological activity. The competence of the biological potential of silver (I) complexes is essentially influenced by the type of ligands bound to the metal centre. The presence of lipophilic groups such as alkyl chains on the NHC ligand enhances the lipophilic nature of the silver complex, which helps it penetrate the cell wall easily. The complexes have shown antibacterial activity to different extents, according to the type of ligand.

Benzimidazolium salts (**2a-j**) and Ag(I) complexes (**3a-j**) were tested against bacterial strains both Gram-positive and negative bacterial. As it was reported in the literatures [50–52], the DMSO did not exhibit any antimicrobial activity. The results are reported [27–31] in **Table 1**. Generally, all the Ag(I) complexes exhibited antibacterial activity against all bacterial strains except, the two compounds **3a** and **3i** were not active against Listeria monocytogenes. While all the benzimidazolium salts (**2a–j**) performed a good antibacterial potential against the test Gram-negative and positive strains and showed bacterial inhibition in the range 14 ± 0.5–36 ± 0.2 mm. There was rarely a difference in the antibacterial activity of benzimidazolium salts (**2a-j**) and Ag(I) complexes (**3a-j**) between all bacterial strains, except that with *Micrococcus luteus* strains, the tested compounds showed better antibacterial potential than others. The observed antibacterial activity of tested complexes is

DOI: http://dx.doi.org/10.5772/intechopen.101950

Microorganisms Compounds	Micrococcus luteus LB 14110	Listeria monocytogenes ATCC 19117	Salmonella Typhimurium ATCC 14028	Staphylococcus aureus ATCC 6538	Pseudomonas aeruginosa
2a	20 ± 0.6	14 ± 0.5	18 ± 0.54	16 ± 0.25	16 ± 0.13
2b	22 ± 0.6	15 ± 0.6	18 ± 0.5	17 ± 0.3	17 ± 0.14
2c	35 ± 0.5	16 ± 0.2	18 ± 0.5	18 ± 0.5	22 ± 0.2
2d	30 ± 0.5	14 ± 0.5	16 ± 0.10	18 ± 0.11	16 ± 0.19
2e	25 ± 0.33	22 ± 0.5	18 ± 0.5	18 ± 0.18	20 ± 0.45
2f	36 ± 0.2	16 ± 0.3	18 ± 0.05	20 ± 0.1	20 ± 0.4
2g	28 ± 0.32	16 ± 0.5	22 ± 0.44	18 ± 0.15	22 ± 0.5
2h	30 ± 0.4	16 ± 0.2	16 ± 0.2	20 ± 0.2	18 ± 0.2
2i	30 ± 0.2	22 ± 0.2	22 ± 0.3	22 ± 0.2	20 ± 0.4
2j	34 ± 0.44	22 ± 0.5	22 ± 0.15	22 ± 0.3	20 ± 0.25
3a	20 ± 0.22	—	22 ± 0.22	18 ± 0.05	18 ± 0.22
3b	18 ± 0.2	20 ± 0.2	16 ± 0.3	20 ± 0.2	18 ± 0.2
3c	16 ± 0.2	18 ± 0.3	18 ± 0.22	16 ± 0.0	16 ± 0.5
3d	22 ± 0.2	16 ± 0.2	14 ± 0.2	20 ± 0.2	16 ± 0.2
3e	18 ± 0.2	18 ± 0.22	18 ± 0.33	18 ± 0.23	18 ± 0.22
3f	30 ± 0.4	22 ± 0.7	30 ± 0.4	25 ± 0.2	19 ± 0.17
3g	22 ± 0.3	16 ± 0.4	22 ± 0.4	18 ± 0.2	18 ± 0.2
3h	10 ± 0.4	14 ± 0.5	12 ± 0.10	14 ± 0.15	16 ± 0.10
3i	32 ± 0.32	—	16 ± 0.15	18 ± 0.1	18 ± 0.15
3j	20 ± 0.4	18 ± 0.5	18 ± 0.24	18 ± 0.5	18 ± 0.16

Table 1.
Zone of bacterial inhibition measured in mm of the synthesized salts and silver complexes [27–31].

Microorganism indicator	*Compounds*	*MIC (mg/ml)*
Listeria monocytogenes **ATCC 19117**	***2h***	*1.25*
	2j	*0.625*
	3f	*0.0048*
	Ampicillin	*0.039*
Salmonella **Typhimurium ATCC 14028**	***2h***	*1.25*
	2j	*0.039*
	3f	*0.0024*
	Ampicillin	*0.625*
Micrococcus luteus	***2h***	*0.3125*
	2j	*0.3125*
	3f	*0.0024*
	Ampicillin	*0.0195*

Table 2.
Minimal bacterial inhibitory concentration measured in mg/mL of benzimidazoles salts and Ag(I) complexes [27–31].

comparable to that of our previous silver complexes [53–55]. The MIC values of tested Ag(I) complexes and their starting material against *L. monocytogenes* ATCC 19117, *Salmonella Typhimurium* ATCC 14,028 and *M. luteus* are presented in **Table 2**.

3.1 Minimum inhibitory concentration (MIC) determination

The antimicrobial activity of compounds **2 h**, **2j**, and **3f** has been reported based on MIC values, which are defined as the lowest concentration of an antimicrobial that visibly inhibits bacterial growth after overnight incubation. As shown in **Table 2**, MIC values ranged between 0.0024 and 0.3125 mg mL^{-1} for M.luteus LB 14110. Listeria monocytogenes ATCC 19117 shows the range from 0.0048 to 1.25 mg mL^{-1} and for Salmonella typhimurium ATCC 14028 the MIC values varied between 0.0024 and 1.25 mg mL^{-1}. The Ag complex **3f** showed better activity than ampicillin against Micrococcus luteus as well as for Salmonella Typhimurium with an MIC of 0.0024 mg/mL. Whereas, L. monocytogenes exhibited an MIC value of 0.0048 mg/mL using the same complex. The MICs of the other compounds were in the range tested.

4. Antioxidant activities

The scavenging activity of the synthesized of the NHC precursors [27–31] is in **Figure 3** and Ag(I) complexes with DPPH (1,1-diphenyl- 2-picrylhydrazyl) is represented in **Figure 4**.

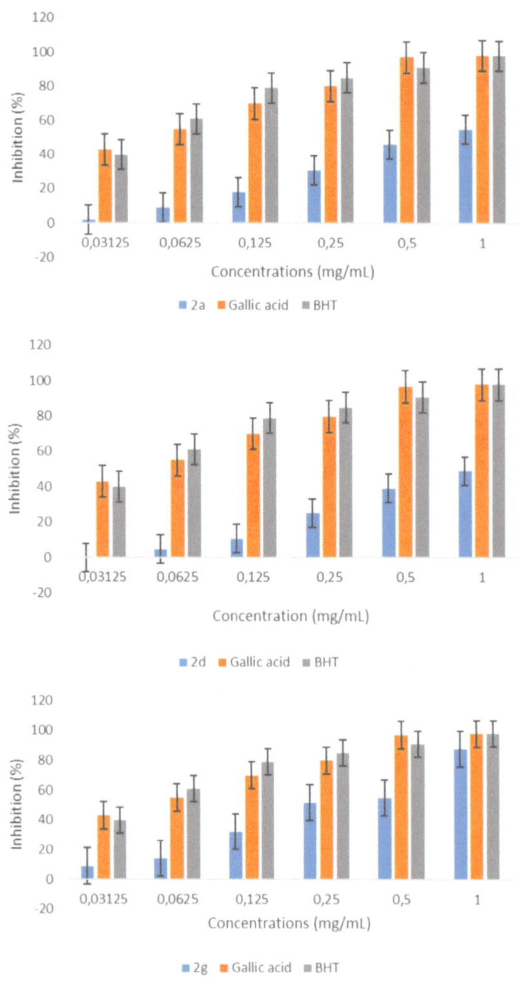

Figure 3.
*DPPH radicals scavenging activity of benzimidazoles salts **2a, 2d, 2g**.*

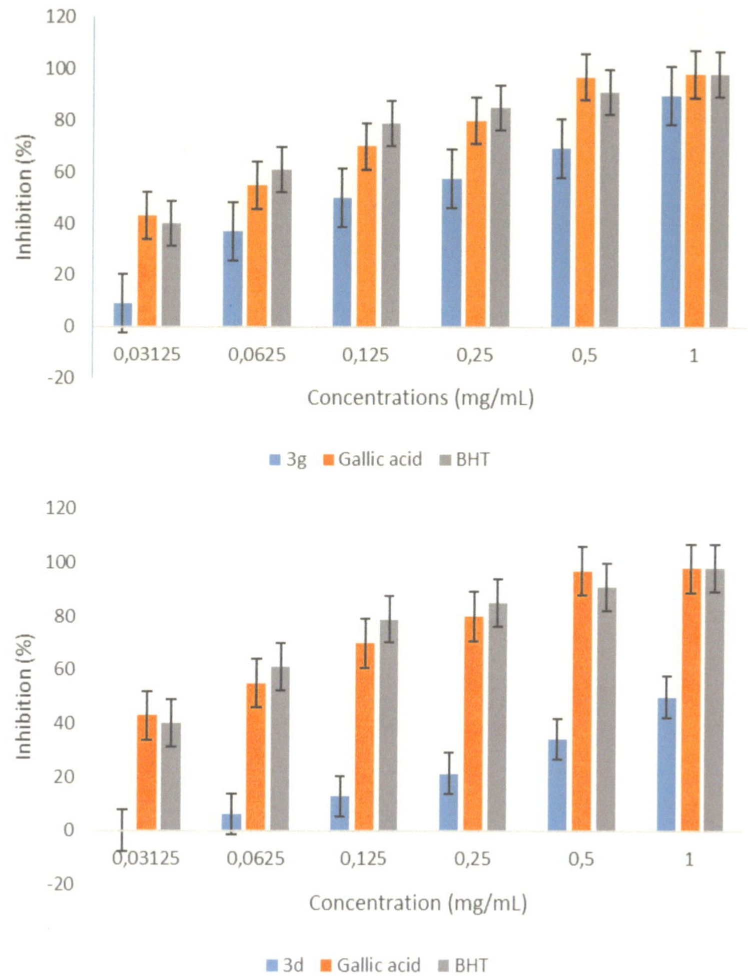

Figure 4.
*DPPH radicals scavenging activity of (Ag-NHC) complexes **3d**, **3g**.*

The antioxidant activities for compounds **2a**, **2d**, **2g**, **3g**, and **3d** are summarized in **Figures 3** and **4**. The results analysis indicated that the antiradical activity profiles obtained from the tested synthetic products **3g** and **3d** had improved and demonstrated antioxidant activity compared to the other products. At a concentration used (0.0625 mg/ml), **2d** showed the lowest free radical activity when compared to both gallic acid and BHT (butylated hydroxytoluene). Similarly, compounds **2a**, **2g** and **3d**, at a concentration of 0.0625 mg/ml, had lower radical activity than gallic acid as well as BHT (butylated hydroxytoluene). **2a**, **2d**, **2g**, **3g** and **3d** revealed significant DPPH activity over synthetic antioxidants at the concentration of 1 mg/ ml.

5. Cytotoxic activities

The anticancer activities of benzimidazole salts **2a-j** and Ag(I) complexes **3a-j** were investigated against breast cancer MCF-7, MDA-MB-231 cells. The results are listed in **Table 3**. The cytotoxicity of **3i** and **3f** was significantly higher against MCF7 cells as shown in **Table 3** with IC50 values of 0.68 and 0.6 mg/ml, respectively, than its activity against MDA-MB-231 cells. Additionally, compound

benzimidazoles salts 2a-j and Ag(I) complexes 3a-j	Anticancer activity LC50 in mg/ml MCF7 MDA-MB-231	
3a	MCF7	MDA-MB-231
3b	4.2 ± 3.6	2.5 ± 4.3
3c	3.1 ± 3.1	2.6 ± 5.9
3d	1.7 ± 3.1	16 ± 2.8
3e	4.3 ± 1.8	0.0 ± 00
3f	0.68 ± 3.2	1.93 ± 2.6
3g	1.3 ± 4.1	3.3 ± 2.9
3h	2.0 ± 3.2	2.8 ± 2.9
3i	0.61 ± 3.1	1.95 ± 2.5
3j	1.3 ± 4.1	3.4 ± 2.9
2a	2.0 ± 3.2	2.7 ± 2.8
2b	NA	NA
2c	3.1 ± 5.9	6.3 ± 3.2
2d	NA	NA
2e	0.6 ± 2.9	3.1 ± 5.9
2f	Higher than 100 mg/ml	Higher than 100 mg/ml
2g	Higher than 100 mg/ml	Higher than 100 mg/ml
2h	Higher than 100 mg/ml	Higher than 100 mg/ml
2i	Higher than 100 mg/ml	Higher than 100 mg/ml
2j	Higher than 100 mg/ml	Higher than 100 mg/ml
Tetracyclinea	NT	NT

Values are mean value ± standard deviation of three different replicates. The concentration was 30 mg, NT: not tested, NA: not active.

Table 3.
*Anticancer activities of synthesized benzimidazoles salts **2a-j** and Ag(I) complexes **3a-j** [27–31].*

3j exhibited cytotoxicity towards MCF7 and MDA-MB-231 cells equal to 2.3 and 3.4 mg/ml. whereas compounds **2a** and **2d** were not active against MCF7 and MDA-MB-231. The compounds **2f-j** had showed IC50 values higher than 100 mg/ml.

On the other hand, benzimidazolium salts (**4a-4j**) have been synthesized following our previous work [56, 57] (**Figure 5**). The ^{1}H NMR spectra of the benzimidazolium salts (**4a-j**) showed an acid proton H_2 which appeared as a typical singlet at 12.02, 11.80, 12.02, 11.77, 11.61, 11.79, 12.15, 12.27, 11.46 and 11.26 ppm, respectively.

The protons of the aromatic group on benzimidazolium salts (**4a-4j**) were identified in the range of 6.30–8.02 ppm. The $H_{2'}$ protons of the isobutyl group were seen as heptate in the range between 2.25 and 2.44 ppm. The signals resonated between 0.98 and 1.04 are assigned to protons of isobutyl group H_{ab} on benzimidazolium salts (**4a-4j**). Further evidence for the formation of benzimidazolium salts (**4a-4j**) is provided by the peak of C2 of the carbons as typical singlets in the range 144.1–144.5 ppm. The ^{13}C NMR spectra showed also aromatic carbons of benzimidazolium salts (**4a-4j**) in the range of 105.8–153.8 ppm. The terminal carbons C_{ab} of the isobutyl group of all benzimidazolium salts (**4a-4j**) showed peaks in the

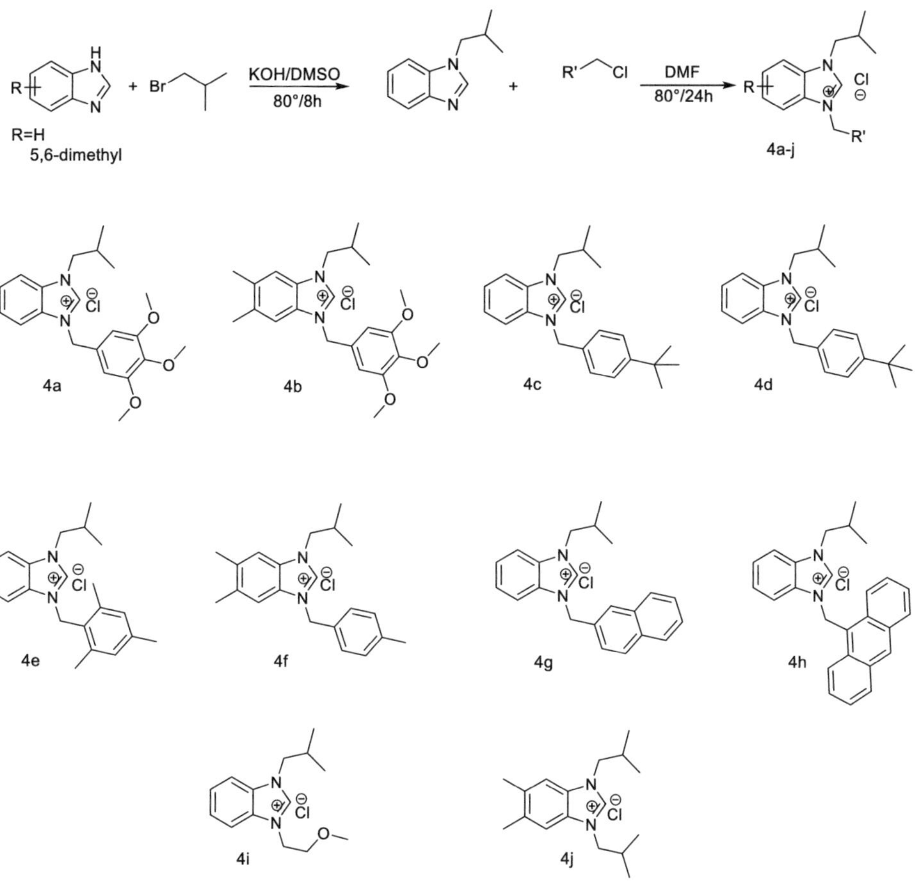

Figure 5.
*Synthesis of benzimidazoles salts **4a-j**.*

region 19.3–19.9 ppm. While the carbons $C_{2'}$ of the isobutyl group were identified between 28.6–28.9 ppm. These values are consistent with those in the corresponding literature [58].

The synthesis of Ag(I) complexes was performed in the absence of light. The reaction is carried out between benzimidazolium salt with 1 equiv. Ag_2O in dichloromethane at room temperature. The Ag(I) complex was produced as a crystalline solid (**Figure 6**). The reaction was monitored by 1H NMR spectroscopy in δ-CDCl3 and demonstrated that the benzimidazolium salts were fully converted to silver complexes in moderate yields (72–93%).

The Ag(I) complexes are stable in air and moisture with high solubility in polar solvents. The formation of the silver carbene complexes was proved by the absence of an NCHN proton peak in their 1H NMR spectra, which confirms the complete conversion to Ag(I) complexes (**5a-5j**).

The successful formation of the silver carbene complexes was also indicated by the presence of the characteristic carbon (NCHN) signals in the bottom region of the field in comparison with those of the corresponding benzimidazolium salts (**4a-4j**). For example, it was observed at 186.7 ppm for complex **5j**. However, the rest of the carbon signal for the rest of the complexes was not observed. These values are in agreement with reported by Asekunowo et al. [59, 60] who have reported the synthesis of a series of monocarbon silver halides [R2NHC]-AgCl and demonstrated the effect of halide ions and solvent on the structural formulas of Ag(I) complexes. In addition, the formation of the Ag(I) complexes (**5a-5j**) was verified by the IR

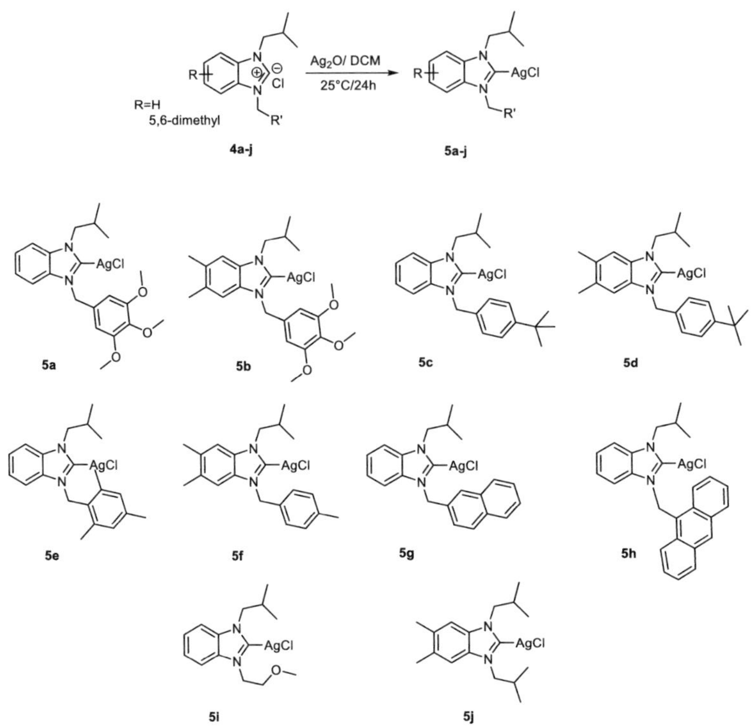

Figure 6.
*Synthesis of Ag-NHC **4a-j**.*

spectra, which showed vibrations of the CN bond at 1567, 1583, 1450, 1467, 1433, 1437, 1450, 1433, 1600 cm^{-1}, respectively.

6. Biological, cytotoxic and antibacterial activities

All the synthesized benzimidazolium salts (**4a-4j**) and their corresponding Ag(I) complexes (**5a-5j**) were investigated for antibacterial against the gram (+)/(−) bacteria. The DMSO did not exhibit any antimicrobial activity as reported earlier [61–73].

All tested compounds exhibited antibacterial activities against all bacteria strains. Compound **5i** was found the most effective in inhibiting the growth of the *Micrococcus luteus* LB 14110. Also, for compounds **5c**, **5 h** and **5f** showed excellent activities against the same bacteria strain. Moreover, NHC precursors (**4a–j**) were less active than corresponding silver complexes (**5a-5j**) against all bacteria strains. The complexes showed an increased antibacterial activity due to the synergistic effect that increases the lipophilicity of the complexes, which facilitates the penetration of the complexes through the cell's membrane.

6.1 MIC determination

The MIC values of tested silver complexes and their starting material against *Listeria monocytogenes* ATCC 19117, *Salmonella* Typhimurium ATCC 14028 and *M. luteus* are presented in **Table 4**.

The antimicrobial activity of compounds **4h**, **4i** and **5f** was also reported in terms of the MIC values, defined as the lowest concentration of an antimicrobial that visibly inhibits the growth of the bacteria after overnight incubation.

Microorganism indicator	Compounds	MIC (mg/ml)
Listeria monocytogenes **ATCC 19117**	**4h**	1.25
	4j	0.635
	5f	0.0058
	Ampicillin	0.049
Salmonella **Typhimurium ATCC 14028**	**4h**	1.26
	4j	0.041
	5f	0.0034
	Ampicillin	0.635
Micrococcus luteus	**4h**	0.3225
	4j	0.3125
	5f	0.0034
	Ampicillin	0.0195

Table 4.
Minimal bacterial inhibitory concentration (MIC) of benzimidazoles salts and Ag(I) complexes [27–31].

As shown in **Table 4**, Silver complex **5f** showed better activity than ampicillin against *L. monocytogenes*, *Salmonella Typhimurium* and *M. luteus* with an MIC of 0.0058, 0.0034, and 0.0034 mg mL – 1, respectively. The NHC precursor **4i** gave a good result with an MIC of 0.041 $mgmL^{-1}$ against *Salmonella Typhimurium.* The other compound performed poorly.

7. Cytotoxic activities

Salts (**4a-j**) and Ag(I) complexes (**5a-j**) were screened for their in vitro anticancer activities on human cancer cell lines MCF7 and MDA-MB-231 using the MTT test. The results are given in **Table 5.**

	Anticancer activity IC50 / (µg mL–1)	
Compounds	**MCF7**	**MDA-MB-231**
5a	4.2 ± 3.5	4.3 ± 3.3
5b	4.1 ± 3.6	2.6 ± 4.3
5c	3.2 ± 3.1	2.7 ± 5.9
5d	1.8 ± 3.1	15 ± 2.8
5e	4.2 ± 1.8	0.0 ± 00
5f	0.69 ± 3.2	1.94 ± 2.6
5g	1.4 ± 4.1	3.4 ± 2.9
5h	2.1 ± 3.2	2.7 ± 2.9
5i	0.63 ± 3.1	1.96 ± 2.5
5j	1.4 ± 4.1	3.5 ± 2.9
4a	2.1 ± 3.2	2.8 ± 2.8
4b	NA	NA

	Anticancer activity IC50 / (μg mL–1)	
Compounds	**MCF7**	**MDA-MB-231**
4c	3.2 ± 5.9	6.2 ± 3.2
4d	NA	5.2 ± 3.1
4e	0.6 ± 2.9	3.1 ± 5.9
4f	> 100	> 100
4g	> 100	> 100
4h	> 100	> 100
4i	> 100	> 100
4j	> 100	> 100
Tetracyclina	**NT**	NT

The concentration was 30 μg. NA: not active; IC_{50}: half maximal inhibitory concentration; MCF7 and MDA-MB-231: human cancer cell lines; NT: not tested. Values are mean value ± standard deviation of three different replicates.

Table 5.
Anticancer activity of synthesized of benzimidazoles salts [27–31] ***4a-4j*** *and Ag(I) complexes* ***5a-5j****.*

The cytotoxicity of **5i** and **5f** was higher in MCF7 with half-maximal inhibitory concentration (IC50) values of 0.63 and 0.69 μg mL^{-1}, respectively, as compared to their activity on MDA-MB-231 cells. Complexes **5j** and **4a** were also determined to be toxic towards MCF7 and MDA-MB-231 with values of (IC50) 2.1 and 2.8 μg mL – 1 respectively. While, the compound 4d was inactive against MCF7.

8. Conclusions

In summary, a series of Ag(I) complexes were synthesized and characterized using different spectroscopic and analytical techniques. Antimicrobial properties of all Ag(I) complexes were evaluated against four Gram-negative, three Gram-positive bacterial strains and two fungal strains. New silver complexes showed high antibacterial activity compared with the precursors against gram (+)/(−) bacteria and fungi strains. Various substituents on nitrogen atoms have a different effect on antimicrobial activity. In addition, the Ag(I) complexes **5i** and **5f** showed good antitumor activity against MDA-MB-231, and MCF-7 cell lines. Moreover, further studies focused on the synthesis of (benz)imidazol-2-ylidene-based silver-complexes and their medical applications as potential metallopharmaceutical agents are currently underway by our research group.

Author details

Ichraf Slimani[1], Khaireddine Dridi[2,3], Ismail Özdemir[4,5], Nevin Gürbüz[4,5] and Naceur Hamdi[1,2*]

1 Research Laboratory of Environmental Sciences and Technologies (LR16ES09), Higher Institute of Environmental Sciences and Technology, University of Carthage, Hammam-Lif, Tunisia

2 Department of Chemistry, College of science, Qassim University, Buraidah, Saudi Arabia

3 Laboratoire de Synthèse Organique Sélective et Hétérocyclique, Faculté des Sciences de Tunis, Université de Tunis El Manar, Tunis, Tunisia

4 Catalysis Research and Application Center, İnönü University, Malatya, Turkey

5 Faculty of Science and Art, Department of Chemistry, İnönü University, Malatya, Turkey

*Address all correspondence to: naceur.hamdi@isste.rnu.tn

References

[1] Arduengo AJ III, Harlow RL, Kline M. A stable crystalline carbene. Journal of the American Chemical Society. 1991;**113**(1):361-363

[2] Benhamou L, Chardon E, Lavigne G, Bellemine Laponnaz S, Cesar V. Synthetic routes to N-heterocyclic carbene precursors. Chemical Reviews. 2011;**111**(4):2705-2733

[3] Díez-Gonzalez S, Marion N, Nolan SP. N-heterocyclic carbenes in late transition metal catalysis. Chemical Reviews. 2009;**109**(8):3612-3676

[4] Enders D, Niemeier O, Henseler A. Organocatalysis by N-heterocyclic carbenes. Chemical Reviews. 2007;**107**(12):5606-5655

[5] Fortman GC, Nolan SP. N-Heterocyclic carbene (NHC) ligands and palladium in homogeneous cross-coupling catalysis: A perfect union. Chemical Society Reviews. 2011;**40**(10):5151-5169

[6] Hopkinson MN, Richter C, Schedler M, Glorius F. An overview of N-heterocyclic carbenes. Nature. 2014;**510**:((7506)485-496

[7] Peris E, Crabtree RH. Recent homogeneous catalytic applications of chelate and pincer N-heterocyclic carbenes. Coordination Chemistry Reviews. 2004;**248**(21-24):2239-2246

[8] Małecki P, Gajda K, Ablialimov O, Malinska M, Gajda R, Wozniak K, et al. Hoveyda–Grubbs-type precatalysts with unsymmetrical N-heterocyclic carbenes as effective catalysts in olefin metathesis. Organometallics. 2017;**36**(11):2153-2166

[9] Meng G, Szostak M. Palladium/NHC (NHC= N-heterocyclic carbene)-catalyzed B-alkyl suzuki cross-coupling of amides by selective N–C bond cleavage. Organic Letters. 2018;**20**(21):6789-6793

[10] Tian X, Lin J, Zou S, Lv J, Huang Q, Zhu J, et al. [Pd (IPr* R)(acac) Cl]: Efficient bulky Pd-NHC catalyst for Buchwald-Hartwig CN cross-coupling reaction. Organometallic Chemistry. 2018;**861**:125-130

[11] Karataş MO, Günal S, Mansur A, Alıcı B, Özdemir İ. Catechol-bearing imidazolium and benzimidazolium chlorides as promising antimicrobial agents. Archiv der Pharmazie. 2020;**353**(6):e2000013

[12] Shi S, Nolan SP, Szostak M. Well-defined palladium (II)–NHC precatalysts for cross-coupling reactions of amides and esters by selective N–C/O–C cleavage. Accounts of Chemical Research. 2018;**51**(10):2589-2599

[13] Domyati D, Latifi R, Tahsini L. Sonogashira-type cross-coupling reactions catalyzed by copper complexes of pincer N-heterocyclic carbenes. Journal of Organometallic Chemistry. 2018;**860**:98-105

[14] Karataş MO, Giuseppe AD, Passarelli V, Alıcı B, Pérez-Torrente JJ, Oro LA, et al. Pentacoordinated rhodium (I) complexes supported by coumarin-functionalized N-heterocyclic carbene ligands. Organometallics. 2018;**37**(2):191-202

[15] Karataş MO. Cycloheptyl substituted N-heterocyclic carbene PEPPSI-type palladium complexes with different N-coordinated ligands: involvement in Suzuki-Miyaura reaction. Journal of Organometallic Chemistry. 2019;**899**:120906

[16] Kaloğlu M, Kaloğlu N, Günal S, Özdemir İ. Synthesis of N-heterocyclic

carbene-based silver complexes and their antimicrobial properties against bacteria and fungi. Journal of Coordination Chemistry. 2021:1-17

[17] Özdemir İ, Demir Düşünceli S, Kaloğlu N, Achard M, Bruneau C. Synthesis of ruthenium N-heterocyclic carbene complexes and their catalytic activity for β-alkylation of tertiary cyclic amines. Journal of Organometallic Chemistry. 2015;**799**:311-315

[18] Şahin Z, Gürbüz N, Özdemir İ, Şahin O, Büyükgüngör O, Achard M, et al. Synthesis of ruthenium N-heterocyclic carbene complexes and their catalytic activity for β-alkylation of tertiary cyclic amines. Organometallics. 2015;**34**(11):2296-2304

[19] Şahin N, Özdemir N, Gürbüz N, Özdemir İ. Novel N-Alkylbenzimidazole-Ruthenium (II) complexes: Synthesis and catalytic activity of N-alkylating reaction under solvent-free medium. Applied Organometallic Chemistry. 2019;**33**(2): e4704

[20] Kaloğlu WJ, Kaloğlu M, Tahir MN, Arıcı C, Bruneau C, Doucet H, et al. Synthesis of N-heterocyclic carbene-palladium-PEPPSI complexes and their catalytic activity in the direct CH bond activation. Journal of Organometallic Chemistry. 2018;**867**:404-412

[21] Lin JCY, Huang RTW, Lee CS, Bhattacharyya A, Hwang WS, Lin IJB. Coinage metal- N-heterocyclic carbene complexes. Chemical Reviews. 2009;**109**(8):3561-3598

[22] Mnasri A, Mejri A, Al-Hazmy SM, Arfaoui Y, Özdemir İ, Gürbüz N, et al. Silver–N-heterocyclic carbene complexes-catalyzed multicomponent reactions: Synthesis, spectroscopic characterization, density functional theory calculations, and antibacterial study. Archiv der Pharmazie. 2021: e2100111

[23] Achar G, Ramya VC, Upendranath K, Budagumpi S. Coumarin-tethered (benz) imidazolium salts and their silver (I) N-heterocyclic carbene complexes: Synthesis, characterization, crystal structure and antibacterial studies. Applied Organometallic Chemistry. 2017;**31**(11):e3770

[24] Kızrak Ü, Çiftçi O, Özdemir İ, Gürbüz N, Düşünceli D, Kaloğlu M, et al. Amine-fnctionalized silver and gold N-heterocyclic carbene complexes: Synthesis, characterization and antitumor properties. Journal of Organometallic Chemistry. 2019;**882**:26-32

[25] Hussaini SY, Haque RA, Razali MR. Recent progress in silver (I)-, gold (I)/(III)-and palladium (II)-N-heterocyclic carbene complexes: A review towards biological perspectives. Journal of Organometallic Chemistry. 2019;**882**:96-111

[26] Mora M, Gimeno MC, Visbal R. Recent advances in gold–NHC complexes with biological properties. Chemical Society Reviews. 2019;**48**(2): 447-462

[27] Slimani I, Mansour L, Abutaha N, Harrath AH, Al-Tamimi J, Gürbüz N, et al. Synthesis, structural characterization of silver (I)-NHC complexes and their antimicrobial, antioxidant and antitumor activities. Journal of King Saud University. 2020;**32**(2):1544-1554

[28] Nayak S, Gaonkar SL, Musad EA, Dawsar AMA. 1, 3, 4-Oxadiazole-containing hybrids as potential anticancer agents: Recent developments, mechanism of action and structure-activity relationships. Journal of Saudi Chemical Society. 2021;**25**(8):101284

[29] Liang X, Luan S, Yin Z, He M, He C, Yin L, et al. Recent advances in the medical use of silver complex. European Journal of Medicinal Chemistry. 2018;**157**:62-80

[30] Asekunowo PO, Haque RA, Razali MR, Avicor SW, Wajidi MFF. Synthesis and characterization of nitrile functionalized silver (I)-N-heterocyclic carbene complexes: DNA binding, cleavage studies, antibacterial properties and mosquitocidal activity against the dengue vector, Aedes albopictus. European Journal of Medicinal Chemistry. 2018;**150**:601-615

[31] Habib A, Iqbal MA, Bhatti HN, Shahid M. Effect of ring substitution on synthesis of benzimidazolium salts and their silver (I) complexes: Characterization, electrochemical studies and evaluation of anticancer potential. Transition Metal Chemistry. 2019;**44**(5):431-443

[32] Russell AD, Hugo WB. 7 antimicrobial activity and action of silver. Progress in Medicinal Chemistry. 1994;**31**:351-370

[33] Feng QL, Wu J, Chen GQ, Cui FZ, Kim TN, Kim JO. A mechanistic study of the antibacterial effect of silver ions on Escherichia coli and Staphylococcus aureus. Journal of Biomedical Materials Research. 2000;**52**(4):662-668

[34] Lansdown AB. Microbial multidrug resistance (mdr) and Oligodynamic silver. Journal of Wound Care. 2002;**11**:125-130

[35] Şahin N, Semeril D, Brenner E, Matt D, Özdemir İ, Kaya C, et al. Subtle steric effects in nickel-catalysed Kumada–Tamao–Corriu cross-coupling using resorcinarenyl-imidazolium salts. European Journal of Organic Chemistry. 2013(20):4443-4449

[36] Şahin N, Semeril D, Brenner E, Matt D, Özdemir İ, Kaya C, et al. Resorcinarene-functionalised imidazolium salts as ligand precursors for palladium-catalysed Suzuki–Miyaura cross-couplings. ChemCatChem. 2013;**5**(5):1116-1125

[37] Chinnabattigalla S, Dakoju RK, Gedu SJ. Recent advances on the synthesis of flavans, isoflavans, and neoflavans. Journal of Heterocyclic Chemistry. 2020;**58**(2):415-441. DOI: 10.1002/jhet.4176

[38] Lu Y, Yang J, Wang X, Ma Z, Li S, Liu Z, et al. Research progress in use of traditional Chinese medicine for treatment of spinal cord injury. Biomedicine & Pharmacotherapy. 2020;**127**:110136

[39] Al-Mulla A. A review: Biological importance of heterocyclic compounds. Der Pharma Chemica. 2017;**9**(13):141-147

[40] Jethava DJ, Borad MA, Bhoi MN, Acharya PT, Bhavsar ZA, Patel HD. New dimensions in triazolo [4, 3-a] pyrazine derivatives: The land of opportunity in organic and medicinal chemistry. Arabian Journal of Chemistry. 2020;**13**(12):8532-8591

[41] Rabuffetti M, Bavaro T, Semproli R, Cattaneo G, Massone M, Morelli CF, et al. Synthesis of ribavirin, tecadenoson, and cladribine by enzymatic transglycosylation. Catalysts. 2019;**9**(4):355

[42] Sun Y, Jiang H, Wu W, Zeng W, Wu X. Copper-catalyzed synthesis of substituted benzothiazoles via condensation of 2-aminobenzenethiols with nitriles. Organic Letters. 2013;**15**(7):1598-1601

[43] Tagg JR, McGiven AR. Assay system for bacteriocins. Applied Microbiology 1971;**21**(5):943

[44] National Committee for Clinical Laboratory Standards (NCCLS). M7-A6: Methods for Dilution Antimicrobial Susceptibility Tests for Bacteria that Grow Aerobically. 6th ed. Wayne: NCCLS; 2005

[45] National Committee for Clinical Laboratory Standards (NCCLS). M38-P:

Reference Method for Broth Dilution Antifungal Susceptibility Testing of Filamentous Fungi. Wayne: NCCLS; 1998

[46] Gulluce M, Sahin F, Sokmen M, Ozer H, Daferera D, Sokmen A, et al. Antimicrobial and antioxidant properties of the essential oils and methanol extract from *Mentha longifolia* L. ssp. *longifolia*. Food Chemistry. 2007;**103**(4):1449-1456

[47] Kirby AJ, Schmidt RJ. Antimicrobial and antioxidant properties of the essential oils and methanol extract from *Mentha longifolia* L. ssp. *longifolia*. Journal of Ethnopharmacology 1997;**56**(2):103-108

[48] Mosmann T. Rapid colorimetric assay for cellular growth and survival: application to proliferation and cytotoxicity assays. Journal of Immunological Methods. 1983;**65**(1-2):55-63

[49] Miles AA, Misra SS, Irwin JO. The estimation of the bactericidal power of the blood. Epidemiology & Infection. 1938;**38**(6):732-749

[50] Zonouzi A, Mirzazadeh R, Safavi M, Kabudanian SA, Emami S, Foroumadi A. 2-amino-4-(nitroalkyl)-4h-chromene-3-carbonitriles as new cytotoxic agents. Iranian Journal of Pharmaceutical Research. 2013;**12**(4):679

[51] Asekunowo PO, Haque RA, Razali MR, Budagumpi S. Benzimidazole-based silver (I)–N-heterocyclic carbene complexes as anti-bacterials: Synthesis, crystal structures and nucleic acids interaction studies. Applied Organometallic Chemistry. 2015;**29**(3):126-137

[52] De Fremont P, Scott NM, Stevens ED, Rammnial T, Lightbody OC, Macdonald CLB, et al. Synthesis of well-defined N-heterocyclic carbene silver (I) complexes. Organometallics. 2005;**24**(26): 6301-6309

[53] Chiu PL, Chen CY, Lee CC, Hsieh MH, Chuang CH, Lee HM. Structural variations in novel silver (I) complexes with bitopic pyrazole/N-heterocyclic carbene ligands. Inorganic Chemistry. 2006;**45**(6):2520-2530

[54] Slimani I, Chakchouk-Mtibaa A, Mellouli L, Mansour L, Ozdemir I, et al. Novel N-heterocyclic carbene silver (I) complexes: Synthesis, structural characterization, antimicrobial and cytotoxicity potential studies. Journal of the Brazilian Chemical Society. 2020;**31**:2058-2070

[55] Liu W, Gust R. Metal N-heterocyclic carbene complexes as potential antitumor metallodrugs. Chemical Society Reviews 2013;42(2):755-773.

[56] Lansdown ABG. Silver I: its antibacterial properties and mechanism of action. Journal of Wound Care. 2014;**11**(4):125-130

[57] Slimani I, Mansour L, Abutaha N, Halim Harrath A, Al-Tamimi J, Gürbüz N, et al. Synthesis, structural characterization of silver (I)-NHC complexes and their antimicrobial, antioxidant and antitumor activities. Journal of King Saud University-Science. 2020;**32**(2):1544-1554

[58] Chen W, Liu F. Synthesis and characterization of oligomeric and polymeric silver-imidazol-2-ylidene iodide complexes. Journal of Organometallic Chemistry. 2003;**673**(1-2):5-12

[59] Lin IJ, Vasam CS. Silver (I) N-heterocyclic carbenes. Comments on Inorganic Chemistry. 2004;**25**(3-4): 75-129

[60] Asekunowo PO, Rosenani AH, Mohd RR. Reviews in Inorganic Chemistry. 2017;**37**:29-50

[61] Hayes JM, Viciano M, Peris E, Ujaque G, Lledós A. Mechanism of

formation of silver N-heterocyclic carbenes using silver oxide: A theoretical study. Organometallics. 2007;**26**(25):6170-6183

[62] Slawson RM, Van Dyke MI, Lee H, Trevors JT. Germanium and silver resistance, accumulation, and toxicity in microorganisms. Plasmid. 1992;**27**(1): 72-79

[63] Zhao G, Stevens SE. Multiple parameters for the comprehensive evaluation of the susceptibility of *Escherichia coli* to the silver ion. Biometals. 1998;**11**(1):27-32

[64] Achar G, Agarwal P, Brinda KN, Małecki JG, Keri RS, Budagumpi S. Ether and coumarin–functionalized (benz) imidazolium salts and their silver (I)–N–heterocyclic carbene complexes: Synthesis, characterization, crystal structures and antimicrobial studies. Journal of Organometallic Chemistry. 2018;**854**:64-75

[65] Achar G, Shahinia CR, Patil SA, Malecki JG, Pan SH, Lan A, et al. Sterically modulated silver (I) complexes of coumarin substituted benzimidazol–2–ylidenes: Synthesis, crystal structures and evaluation of their antimicrobial and antilung cancer potentials. Journal of Inorganic Biochemistry. 2018;**183**:43-57

[66] Adhikary SD, Jhulki L, Seth S, Kundu A, Bertolasi V, Mitra P, et al. Synthetic strategy of diflurophosphate-bridgedbimetallic N- heterocycliccarbenecomplexes: Synthesis, structures and photoluminescence of picolyl-substituted alkylbenzimidazolylidene ligands. Inorganica Chimica Acta. 2012;**384**:239-246

[67] Sánchez OS, Gonzalez S, Higuera-Padilla AR, Leon Y, Coll D, Fernandez M, et al. Remarkable in vitro anti-HIV activity of new silver(I)e and gold(I)-N- heterocyclic carbene complexes. Synthesis, DNA binding and biological evaluation. Polyhedron. 2016;**110**:14-23

[68] Arduengo AJ III, Dias HVR, Calabrese JC, Davidson F. Homoleptic carbine silver(I) and carbine-copper(I) complexes. Organometallics. 1993;**12**(9):3405-3409

[69] Arduengo AJ III, Harlow RL. Stable crystalline carbine. Journal of the American Chemical Society. 1991;**113**(1):361-363

[70] Balcan S, Balcan M, Cetinkaya B. Poly(l-lactide) initiated by silver N- heterocyclic carbene complexes: Synthesis, characterization and properties. Polymer Bulletin. 2013;**70**(12):3475-3485

[71] Baquero EA, Silbestri GF, Gomez-sal P, Flores JC, Jesús E. Sulfonated water-soluble N-heterocyclic carbine silver(I) complexes: Behavior in aqueous medium and as NHC-transfer agent to platinum(II). Organometallics. 2013;**32**(9):2814-2826

[72] Boubakri L, Mansour L, Harrath AH, Özdemir I, Yasar S, Hamdi N. *N*-heterocyclic carbene-Pd (II)-PPh3 complexes a new highly efficient catalyst system for the Sonogashira cross-coupling reaction: Synthesis, characterization and biological activities. Journal of Coordination Chemistry. 2018;**71**(2):183-199

[73] Cavallo L, Correa A, Costabile C, Jacobsen H. Steric and electronic effects in the bonding of N-heterocyclic ligands to transition metals. Journal of Organometallic Chemistry. 2005;**690** (24-25):5407-5413

Chapter 7

Imidazolium-Based N-Heterocyclic Carbenes (NHCs) and Metal-Mediated Catalysis

Vittal B. Gudimetla, Bony P. Joy and Sudeep Paul

Abstract

The journey of "carbenes" is more than a century old. It began with a curiosity to understand a then less familiar carbon moiety in its divalent state. It reached an important milestone in the form of 1,3-imidazolium-based N-heterocyclic carbenes (NHCs), where the quest for bottleable carbenes was achieved through simple and elegant synthetic routes. The properties of these carbenes were finely tunable through the steric and electronic factors *via* chemical modifications. Thus, it became one of the unique and extensively studied ligands for its properties and applications. This chapter first briefs about structural details of NHCs and different synthetic routes for the preparation of imidazolium-based NHC precursors. The later section focuses on various methods for characterizing the steric and electronic properties of these ligands and their metal intermediates, which are crucial for developing efficient catalytic processes. Finally, the chapter concludes with NHC-metal-mediated catalytic applications and its immediate challenges.

Keywords: N-heterocyclic carbenes, imidazolium salts, NHC-metal-mediated catalysis, steric and electronic properties, characterization of carbenes

1. Introduction

Carbon in its low coordinate divalent state is known as "carbene." It is a well-known transient species in many catalytic reactions. Initially, the carbenes were considered to be very reactive and difficult to isolate. However, the carbenes were first stabilized as metal carbene complexes, which have found numerous applications in many important organometallic reactions. The challenge of isolating a stable metal-free carbene was ended in 1991, through the pioneering works of Arduengo et al. It was for the first time a stable metal-free N-heterocyclic carbene (NHC) was synthesized and isolated in the form of 1,3-di(adamantyl)imidazol-2-ylidene (IAd) by deprotonation of its 1,3-imidazolium salt precursor [1]. Thus, it enabled the easy synthesis of NHCs using simple synthetic routes and thereby extended its scope in synthetic organic chemistry. One such celebrated success is the role of NHCs as ligands in olefin metathesis [2].

1,3-Imidazolium salts are simple and stable precursors of NHCs. They are a good source for *in situ* production of NHCs either by heating or upon reaction with a base. Chemically, these NHCs are nitrogen-containing heterocyclic compounds with a

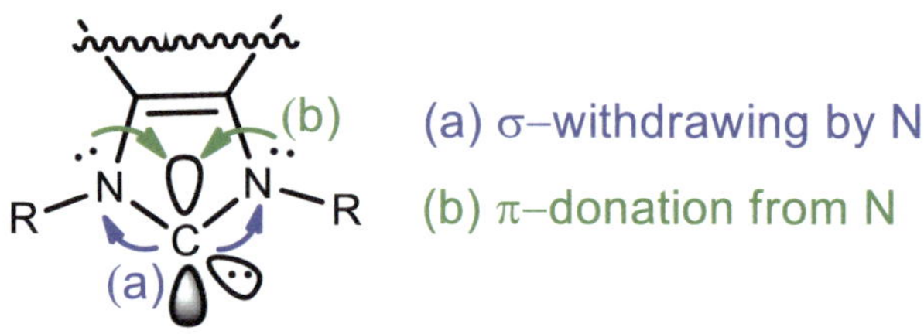

Figure 1.
Schematic representation of σ-withdrawing and π-donating electronic effects through the adjacent nitrogen atoms with the carbene carbon via *inductive and mesomeric effects, respectively.*

divalent carbene center [3]. Due to the available lone pair of electrons on the carbon, the carbene in these NHCs were often compared with phosphine ligands and validated for its effectiveness. The stability of this electron-rich carbene and its unique electronic properties are primarily due to the presence of two adjacent nitrogen atoms, which will have a push-pull electronic influence at the carbene center (**Figure 1**). The effect is synergistic in nature, where the N atom in the NHC withdraws the σ-electrons inductively and donates π-electrons to the imidazolium ring mesomerically. The inductive effect of the N-atom lowers the highest occupied molecular orbital (HOMO) energy level occupied by the *p*-orbital and thereby increases the energy gap between HOMO and lowest unoccupied molecular orbital (LUMO) levels of NHC. Hence, the second electron gets paired in the HOMO level resulting in the singlet form of carbene. Furthermore, the N-atom mesomerically donates a pair of non-bonding electrons from its *p*-orbital and overlaps with the empty *p*-orbitals of the C-atom, which reinforces the stability of the singlet carbene. Moreover, the cyclic bent structure of NHCs forces the carbene to a singlet state, whereas in acyclic carbenes, it exists in the triplet state [3–8]. Singlet NHCs are capable of forming stable and stronger bonds with metals due to their good σ-donor and π-acceptor properties [9, 10], especially with electron-deficient metal atoms they act as good π-donors [11]. These unique electronic properties will equip NHCs to have similar or sometimes better reactivity than phosphines; hence, they are considered to be one of the important ligands in catalysis.

In addition to the unique electronic features, the structural diversity of the NHCs was another important aspect that attracted the attention of chemists. The scope for variation of the ring size from 4 to 7-membered ring structures and the possibilities for substituent variations either at the N-atom or on the C-atom of the ring have tremendously increased the structural diversity of NHCs. Dimer, trimer, and poly NHCs are also known in the literature. Thus, the relatively easy access to structurally diverse NHCs and its flexible metal coordination has resulted in a large number of NHC-metal complexes. Based on the metal coordination to the NHC unit, they have been classified as normal, abnormal, and remote NHCs (**Figure 2**). If it binds to the metal through the C_2 atom, then it is known as normal NHC;

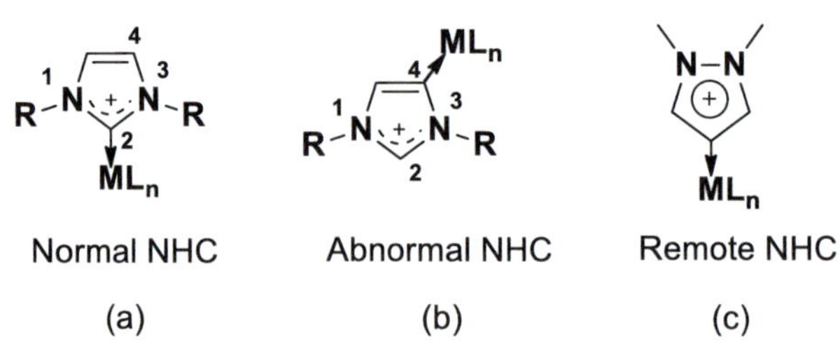

Figure 2.
Representative structures of (a) normal NHC; (b) abnormal NHC; and (c) remote NHC.

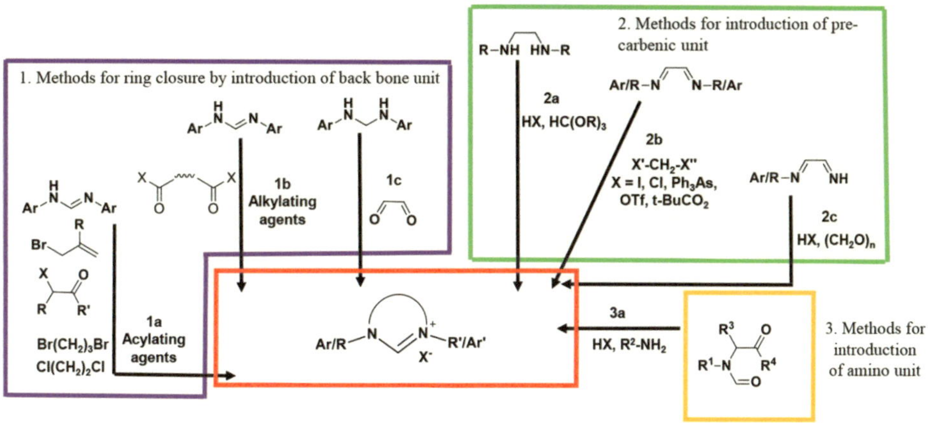

Figure 3.
Different synthetic strategies for making NHC precursors [14].

similarly if it is through the C_4 atom, it is known as abnormal NHC. If there is an absence of a heteroatom in the α-position of the carbenic coordination center, it is known as remote NHC. Compared with normal NHCs, the abnormal NHCs have stronger donor ability and hence have direct impact on its catalytic efficacy [12].

The typical synthesis of NHCs were often accomplished by deprotonation of azolium salts, such as imidazolium, thiazolium, and triazolium. The focus of this chapter is primarily on the imidazolium-based NHCs. In the synthesis of imidazolium-based NHCs, there are three key components, which form the NHC framework. They are (1) carbon backbone, (2) pre-carbenic unit, and (3) amino unit [13]. Different synthetic strategies were evolved based on the construction of these three key components (**Figure 3**). Apart from the conventional synthetic methods, other methods such as electrochemical, microwave, and solvent free synthetic methods were also known [15–17]. Recently, the incorporation of one or more NHC units into a single carbon backbone was also reported, and these multidentate ligands were used to make poly NHC-metal complexes. These poly NHCs are capable of exhibiting different steric and electronic properties than the normal NHCs. Moreover, due to the inherent structural chirality, they are often studied for applications in asymmetric catalysis [18–23].

Considering the scope and potential applications of imidazolium-based NHCs *via* structural refinement, it is essential to understand the factors that will control its properties and catalytic efficiency. Therefore, the following sections elaborates the methods to assess the steric and electronic properties of the NHCs.

2. Methods for assessing electronic and steric parameters of NHCs

2.1 IR spectroscopy: Tolman electronic parameters (TEP)

The most extensively used method for the determination of electronic properties of NHCs is based on the Tolman electronic parameters devised by Tolman in 1977, specifically for phosphines [24]. This technique uses tetrahedral $Ni(CO)_3L$ as model complex [25], where L is the ligand whose donor-acceptor abilities are under investigation. Carbonyl being a good π-acceptor has strong metal-to-carbonyl (d_π-p_π) back bonding. The introduction of ligands (L) with strong electron-donor properties will enhance the electron density around the metal (Ni) center in the $Ni(CO)_3L$ complex.

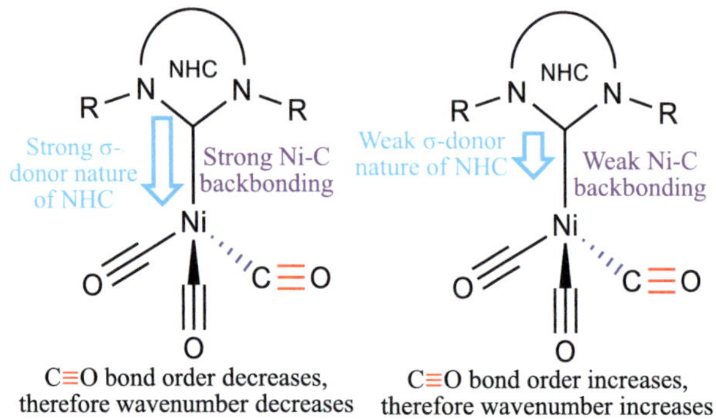

Figure 4.
Determination of TEP using IR spectroscopy [25].

Thus, by comparing the stretching frequencies of the carbonyl (CO) unit, various ligands (L) of interest were analyzed by their σ-donor properties. Ligands (L) with good σ-donor ability increase the metal-to-carbonyl back bonding. It results in stronger metal-ligand (Ni-L) bonding. Thus, it indirectly reduces the bond order of the carbonyl unit, and it will be reflected in the carbonyl IR stretching frequency (**Figure 4**). Nolan et al. were the first to systematically determine the TEP of NHCs using $[Ni(CO)_3(NHC)]$ as model complex. The critical limitation of this method is the preparation of a variety of nickel complexes using different NHC ligands. This synthesis requires the handling of NHCs with low boiling and toxic $Ni(CO)_4$. Moreover, the TEP values are within a small window of 10 cm^{-1} and hence require high-resolution IR spectrophotometers. Any anomalies inherent to the IR measurement technique make this method more challenging. TEP depends on the steric factors since the steric can influence the complexes geometry, thereby the overlapping orbitals affect the TEP value. To overcome few of these shortcomings with respect to synthesis of analytical probes, Crabtree et al. [26] in 2003 reported the use of $[(NHC)M(CO)_2Cl]$, (M = Ir or Rh) complexes as the probe for determining the TEP. In 2008, Nolan et al. reported an extensive study on the Ir complexes for TEP determination of NHCs [27]. Compared with the earlier model complexes, an average of two carbonyl stretching vibration are taken into account due to *cis* arrangement of the carbonyl ligands in the $[(NHC)M(CO)_2Cl]$, (M = Ir or Rh) complex. Thus, the IR spectroscopy was found to be an useful tool to assess the Tolman electronic parameters (TEP). TEP of selected NHCs are given in **Table 1**.

2.2 NMR spectroscopy: σ-donor and π-acceptor abilities of NHCs

NMR spectroscopy, especially ^{13}C and ^{1}H NMR, are the most efficient and reliable techniques to understand the properties of NHCs, the NHC precursors, and

Sl. No.	NHC	TEP (cm^{-1})
1.	IAd	2049.5
2.	IPr	2051.5
3.	SIPr	2052.2
4.	IMes	2050.7
5.	ItBu	2050.1
6.	ICy	2049.6

Table 1.
TEP values for selected NHCs.

related NHC-metal complexes. In general, imidazolium proton (-N-C(H)-N-) in NHCs is acidic due to its connection to the two adjacent electron-withdrawing nitrogen atoms. Hence, this proton appears in the deshielded region of proton NMR, around 8–12 ppm. However, upon deprotonation, the imidazolium proton signal disappears in the proton NMR. The $^{13}C\{^{1}H\}$ NMR spectra of NHCs exhibit the carbene (-N-C-N-) carbon peak close to the highly deshielded region of around 200–250 ppm for free carbene, whereas it is approximately 130–160 ppm for its respective precursor imidazolium salt. Due to the increased shielding effect, the highly shielded carbene carbon upon metal complexation, will undergo an upfield shift. Apart from a diagnostic tool, the NMR technique is helpful to explore various properties of NHCs such as (a) the steric properties; (b) σ-donor ability from ligand to the metal center; and (c) π-back bonding from metal to carbene. Understanding these parameters of NHCs is crucial for controlling the catalyst selectivity, reactivity, and efficiency. Earlier, the Tolman electronic parameters (TEP) obtained from the CO stretching frequencies of metal complexes such as $[Ni(CO)_3L]$ and $[MCl(CO)_2L]$ (M = Rh or Ir) were used to assess the electronic properties. However, this method has limitations due to inherent inaccuracies in IR spectroscopy measurements of metal complexes. Also, the necessity of preparing these complexes under highly inert conditions, the requirement of expensive metals, and the usage of highly toxic CO gas are the important drawbacks of the TEP method. Moreover, the TEP only gives information about the overall electron density around the metal center but not independent σ-donor or π-back bonding abilities.

Huynh et al. reported the unique use of NMR for studying the donor properties of NHCs [28]. This method utilizes the ^{13}C-NMR technique, where the *trans*-$[PdBr_2(^{i}Pr_2\text{-bimy})L]^{n-}$ ($^{i}Pr_2$-bimy = 1, 3-diisopropylbenzimidazolin-2-ylidene; L = NHC ligand under investigation) complex was the spectroscopic probe. In this complex, the presence of ligands (L) with the good σ-donor ability induces a downfield shift of the carbene carbon peak of the NHC (i.e., $^{i}Pr_2$-bimy). Thus, it enabled easy comparison of NHCs with other significant ligands, such as phosphines, amines, and isocyanides. Moreover, this method utilizes easily preparable spectroscopic probes. Based on such carbene carbon chemical shift values, the NHC ligands are arranged on a unique σ-donor scale as shown in **Figure 5**.

Betrand et al. in 2013 reported the determination of π-acceptor ability of NHC ligands using the ^{31}P chemical shift value of carbene-phosphine adducts, which exist in two different resonance forms “I” and “II” (**Figure 6**) [29]. The phosphaalkene form (I) has a P=C bond, and the form (II) has P-C dative bond with P having two lone pairs of electrons. The upfield phosphorus NMR chemical shift associated with P atom in the carbene-phosphine adducts agrees with the electron-rich resonance form (II). It suggests that the carbene carbon prefers a dative bond with phosphorus rather than a π bond. If the carbenes have π-accepting property, it should have reflected through a downfield shift in the ^{31}P NMR.

Therefore, the NMR-based method provides an independent way to determine the π-accepting ability of NHC ligand, independent of its σ-donor abilities, which was one of the limitations of Tolman’s method for determining the electronic properties of NHCs. The arrangement of a few known NHCs according to their π-accepting ability, based on the ^{31}P NMR chemical shift values, are provided in **Figure 6**.

Ganter et al. reported a similar technique in which the ^{77}Se chemical shift of carbene-selenium adducts were used to assess the π-acceptor properties of the NHC ligands [30]. The carbene-selenium adduct exists in two different resonance forms

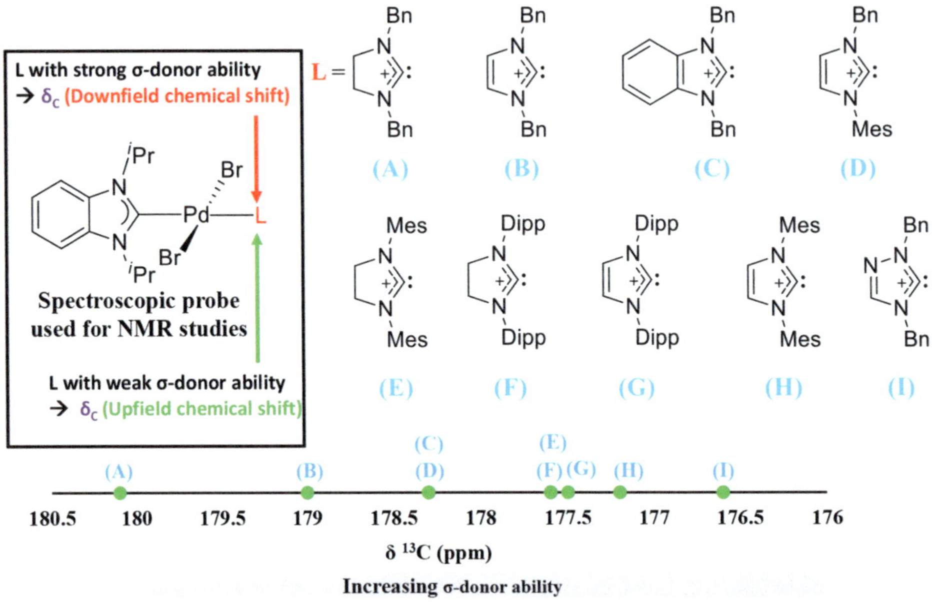

Figure 5.
Arrangement of NHCs based on their σ-donor ability as per the ^{13}C *NMR data, using the spectroscopic probe-*trans*-*$[PdBr_2(^iPr_2\text{-}bimy)L]^{n-}$ *[28].*

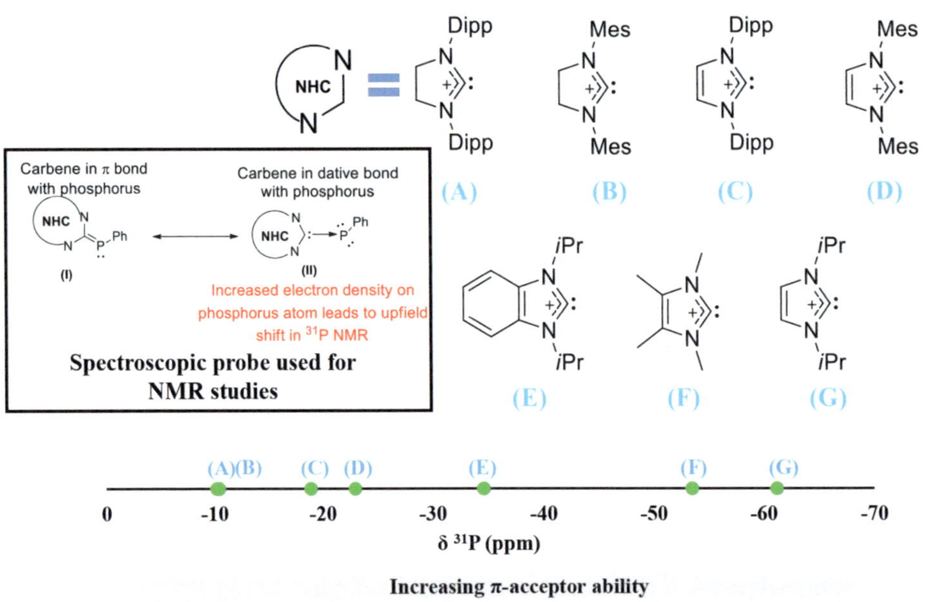

Figure 6.
Arrangement of NHCs according to their π-accepting ability based on the ^{31}P *NMR chemical shift values [29].*

(III) and (IV) (**Figure 7**), similar to that of carbene-phosphine adduct. The increase in the π-accepting property of the NHC is reflected through an increase in the chemical shift value of ^{77}Se NMR. The arrangement of common NHCs according to their π-accepting ability was established based on the ^{77}Se NMR chemical shifts (**Figure 7**). In addition to the π-acceptor ability, the σ-donor abilities of NHCs can also be found using carbon-selenium coupling constant values [31].

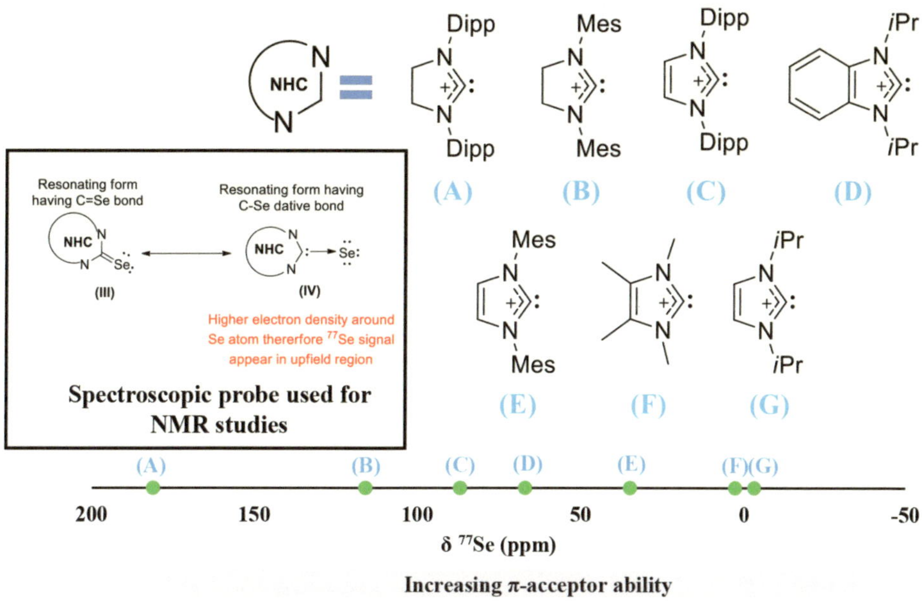

Figure 7.
Arrangement of NHCs according to their π-accepting ability based on the ^{77}Se NMR chemical shift values [30, 31].

Szostak et al. [32] in 2019 reported a simple and straightforward method for the determination of the σ-donor properties of NHCs based on the $C_{carbene}$-H coupling constant ($J_{C\text{-}H}$). The significance of this method in comparison to the earlier reported NMR techniques and TEP determination method is that there is no need for the synthesis of any metal complexes or adducts for the analysis. The ^{1}H NMR is used for the characterization. The magnetic moment of a ^{13}C nuclei couples to a bonded proton through the intervening bonding electrons, and it is represented by the carbon-proton coupling constant ($J_{C\text{-}H}$). This value relies on the probability of finding bonding electrons at the two nuclei involved, that is, C and H. The likelihood of finding an electron at the nucleus of a pure *p*-orbital of carbon is zero, while it has a finite value in the case of an *s*-orbital. Based on this principle, an empirical relationship (i.e., $J_{C\text{-}H} = 500 \times s$) between the C-H coupling constant ($J_{C\text{-}H}$) and the *s*-character of carbon atom (*s*-value = 0.25 (for sp^3-C), 0.33 (for sp^2-C), and 0.50 (for *sp*-C)) was derived, for a 500 MHz ^{1}H NMR instrument. Therefore, carbon with increased "*s*" character is expected to have a large ($J_{C\text{-}H}$) coupling constant.

Thus, a large carbene carbon and proton coupling constant ($J_{C(carbene)\text{-}H}$) value corresponds to a weaker σ-donor property of the NHC ligand. The coupling constants ($J_{C(carbene)\text{-}H}$) were obtained from the ^{13}C satellite peaks of ^{1}H NMR and the ^{1}H coupled ^{13}C NMR spectra. Based on the $J_{C\text{-}H}$, coupling constant values, NHCs have been arranged according to their σ-donor properties (**Figure 8**).

2.3 NMR and UV-Vis spectroscopy: *p*K$_a$ of NHC precursors

Basicity and catalytic reactivity of NHC precursors (azolium salts) are intertwined. Therefore, the *p*K$_a$ studies of NHCs are crucial for understanding the properties of the carbenes. Imidazolium salts upon deprotonation form NHCs (**Figure 9**). The *p*K$_a$ values reflect the ease of imidazolium salts to form NHC precursors. It also provides information about the nucleophilicity of NHCs. Often

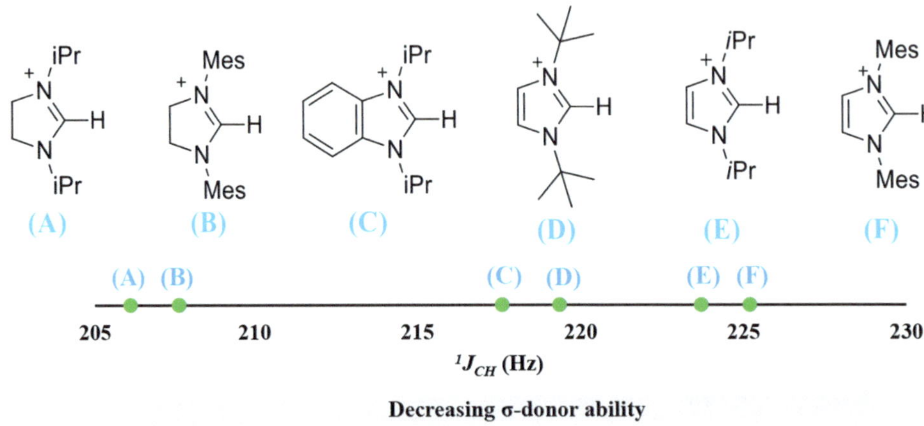

Figure 8.
Arrangement of NHCs according to their σ-donor ability based on the $J_{C(carbene)-H}$ *coupling constant values [32].*

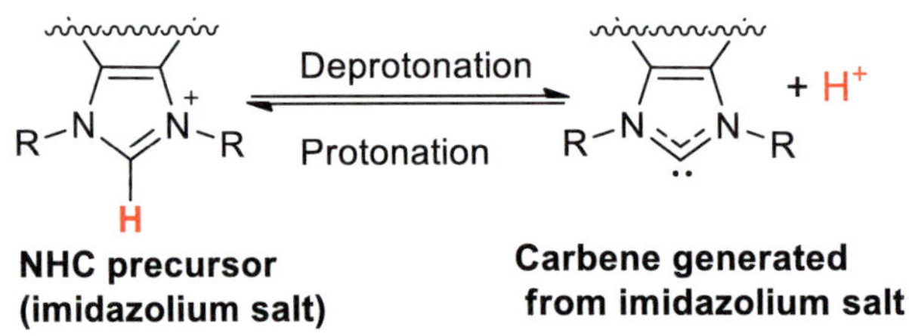

Figure 9.
Deprotonation of imidazolium salts to form N-heterocyclic carbenes.

the NHCs are formed *in situ* and in many NHC-catalyzed reactions, nucleophilic addition is the first and most crucial step. Hence, the details related to nucleophilicity of NHCs have gained considerable interest [33]. Moreover, it will be helpful to draw the correlation between the substituents, nature of the azolium ring, and its catalytic activity. This section highlights the use of NMR and UV-Visible spectroscopy tools for studying the pK_a values of NHC precursors, that is, imidazolium salts.

Alder et al. reported the pK_a value of 1,3-diisopropyl-4,5-dimethylimidazolium cation to be 24.0 in DMSO-*d6* using ^{1}H-NMR spectroscopy [34]. This study uses the NMR technique to assess the deprotonation ability of NHCs from acidic hydrocarbons of known pK_a values (e.g., indene, 9-phenylxanthene, triphenylmethane, etc.) (**Figure 10**). If the NHC deprotonates the hydrocarbon (V), it gives a different proton signal for its corresponding anion (VI). Therefore, based on the proton integration and the equilibrium ratio of (V) and (VI), the pK_a of NHCs can be determined.

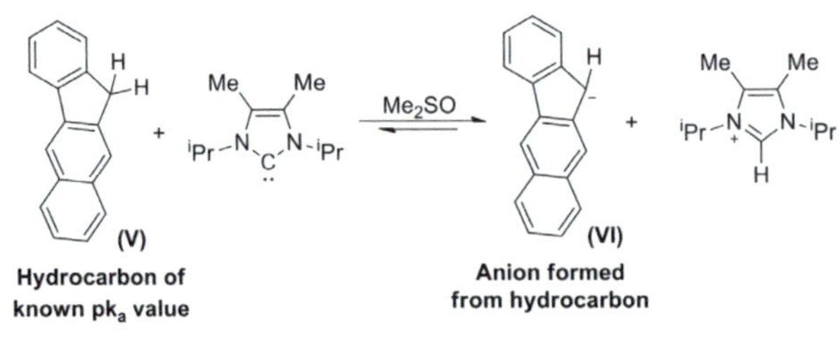

Figure 10.
Deprotonation of hydrocarbon by NHC to form hydrocarbon anion [34].

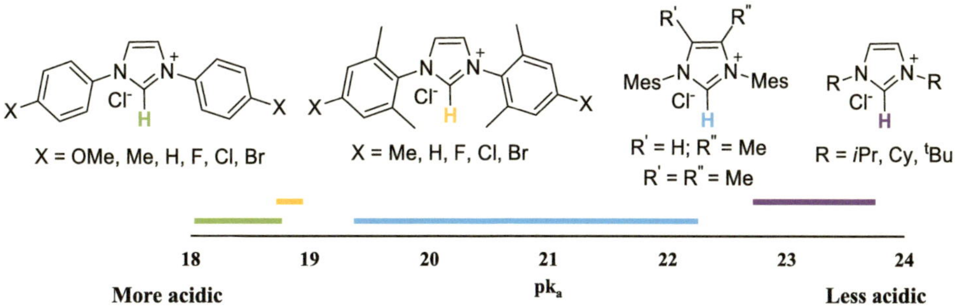

Figure 11.
Arrangement of imidazolium salts based on their pK_a *values in DMSO as determined by Harper et al. using bracketing indicator method [33].*

Streitwieser et al. reported that the anomalies in the determination of pK_a of imidazolium salts, is due to the interaction of acidic solvent (DMSO) with carbene. It can be eliminated through the use of inert solvents such as THF by using bracketing/overlapping indicator method [35]. In this process, the deprotonated form of the fluorene-based indicator (of known pK_a value) shows a significant change in the UV-Vis absorption value upon adding the NHC precursor salt. It helps to monitor the equilibrium and thereby the pK_a can be determined. Chu et al. extended this method for various alkyl imidazolium salts [36]. Harper et al., applied this method to determine the pK_a values of different types of electronically, sterically diverse alkyl and aromatic imidazolium salts (**Figure 11**) [33]. Later the pK_a values of a few imidazolium salts in aqueous medium were also reported by Amyes et al. [37] using deuterium exchange studies. Similar exchange studies by O'Donoghue et al. [38] on triazolium salts facilitated a detailed understanding of the correlation between the NHC precursor salt's cationic structures and its acidity.

2.4 Electrochemical technique: redox potentials and basicity of NHCs

The architecture of NHC allows facile structural modifications, and hence, it provides an opportunity to fine-tune the properties. The reactivity of NHCs is controllable through the structural variations on its cyclic carbon backbone or the introduction of different heteroatoms such as S or O in the heterocyclic ring. The nature of counter anions in the precursor (i.e., imidazolium salts) will also affect the reactivity. Several external factors also contribute to the overall property of NHCs, such as the solvent effects, type of reagents in the reaction mixture, and temperature, etc. Therefore, considering all these factors, the electrochemical techniques will also be helpful for the easy identification of ideal conditions for optimizing the reactivity and efficient catalysis of NHCs. During the electrochemical reduction of imidazolium salt, the imidazolium proton (i.e., C2-H group of the cationic part) $NHCH^+$ is reduced to NHC. In this case, the reactivity mainly depends on the NHC generated from the conjugated acid, that is, $NHCH^+$.

The electrochemical studies of NHCs were first reported by Enders and Simonet et al. using triaryl-triazol-ylidene [39]. The cyclic voltammetry (CV) of imidazolium salts exhibited a reversible reduction, which can be due to single-electron reduction of $NHCH^+$ to NHC (i.e., imidazolium salt to NHC) (**Figure 12**). Recently, Boydston et al. have also reported a systematic study on the redox behavior of a series of azolium salts, including benzothiazolium, thiazolinium, thiazolium, triazolium, imidazolium, and imidazolinium salts [40]. The study demonstrated that N-aryl moiety would be helpful to tune the reduction potential of the

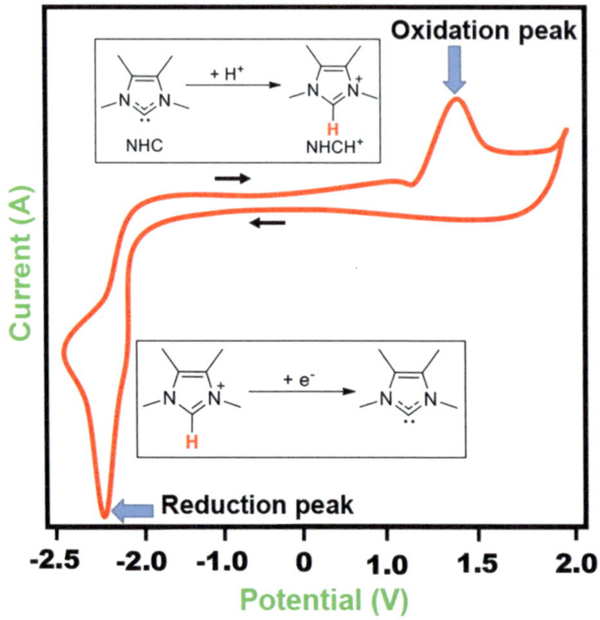

Figure 12.
Schematic CV representation of oxidation and reduction peaks of a model NHC [39–41].

imidazolium ring [41]. The carbene generation usually depends upon the acidity of the $NHCH^+$; hence, with an increase in the reduction potential, the acidity of the $NHCH^+$ also increases [42]. Therefore, based on the electrochemical method, it is possible to choose appropriate NHC precursors [43]. A number of electrochemical studies related to the NHC metal complexes were also reported [44–48], the data of few selective NHC precursors (i.e., azolium salts) and its corresponding reduction potential are summarized from the literature (**Table 2**).

Sl. No.	Compounds	Reduction potential ($E_{1/2}$), (in V) vs SCE
1.	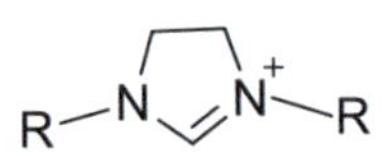 Imidazolinium salt	R= Mes, -2.4 V
2.	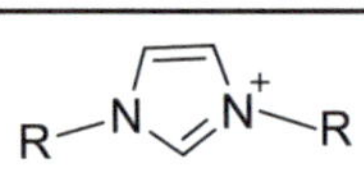 Imidazolium salt	R = 2,6-Dipp, -2.23 V R = Mes, -2.87
3.	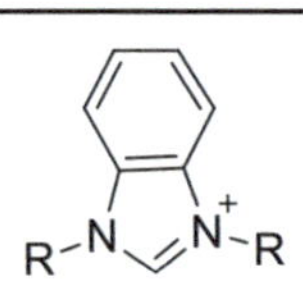 Benzimidazolium salt	R = Me, -2.08 V

Table 2.
Reduction potentials for azolium salts, in V vs. *SCE [40–43].*

The overall electrochemical reduction potential will reflect the acidic nature of the cationic part of the NHC and its nucleophilicity (availability of lone pair). Therefore, the reduction potential data of the azolium salts will be helpful to gain insight into the nucleophilicity of different NHCs and identify the structural moieties that were crucial for a specific function or property. The catalytic efficiency of the NHC ligand primarily depends upon the nucleophilicity of carbene.

3. Steric properties of NHCs

Apart from the electronic properties, it is significant to understand the steric parameters of NHCs, since it plays a crucial role in the reductive elimination step of a catalytic cycle. In case of chiral NHCs, the steric parameters control the enantioselectivity of a catalytic reaction (**Figure 13**).

The earlier attempts to define the steric properties of NHCs were based on the Tolman cone angle [49], which is the solid angle made by an imaginary cone with metal atom at its apex, and ligands are at the outer edges (**Figure 14a**). The cone angle

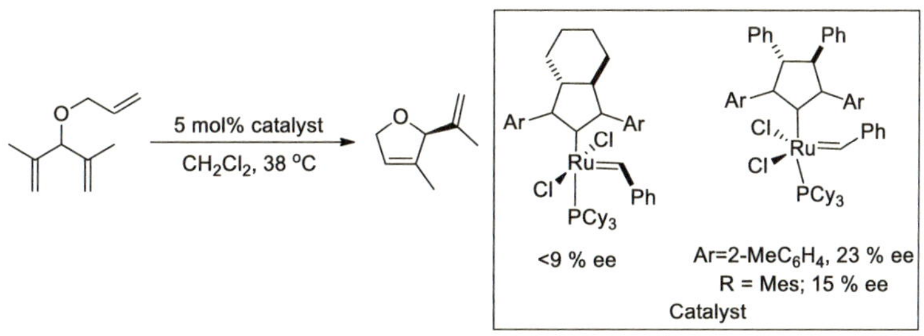

Figure 13.
Effect of carbon backbone and N-substituent modifications on enantioselectivity.

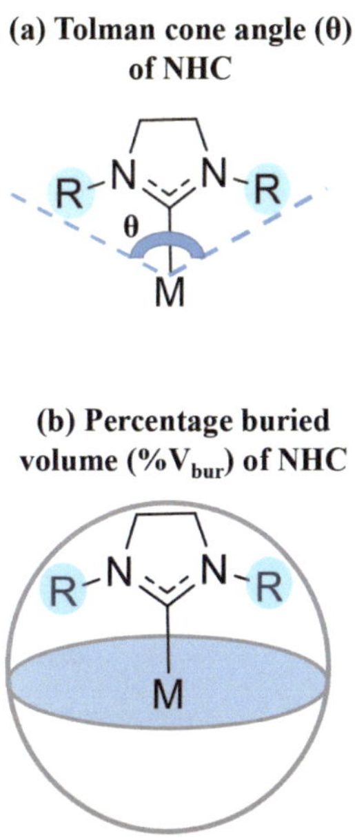

(c) %V_{bur} of selected NHC determined by SambVca

Sl. No.	NHC	% V buried[a]
1.	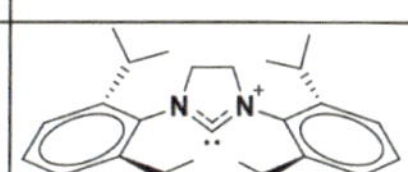	47.0
2.	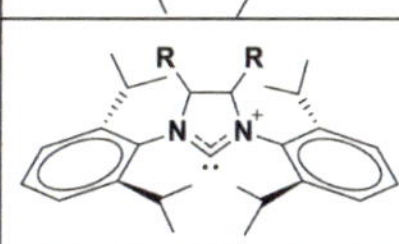	R = H, 45.4 R = Me, 44.4 R = Cl, 44.9
3.	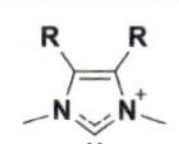	R = H, 26.3 R = Me, 26.2 R = Cl, 26.3
4.	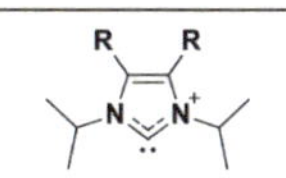	R = H; 27.5 R = Me; 38.5
5.	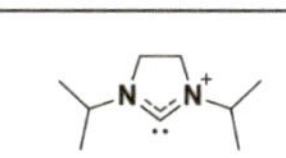	28.2

Figure 14.
(a) Tolman cone angle of NHCs; (b) percentage buried volume of NHC (%V_{bur}); and (c) %V_{bur} for some selected NHCs [50].

can be computationally calculated [51] or obtained from the single-crystal XRD data [52]. In the case of tertiary phosphine ligands, the cone angle was a well-defined concept; however, not in the case of NHCs, where it lacks a predominant three-dimensional symmetry. Thus, the determination of steric parameters, through the Tolman cone angle technique, was inappropriate for NHCs [49]. In order to overcome this, the concept of percentage of buried volume (%V_{bur}) was introduced by Nolan et al. [53], which is the percentage of total volume occupied by the ligand in an imaginary sphere of a well-defined radius and the metal atom residing at the center of the imaginary sphere (**Figure 14b**). This parameter can be determined either from the single-crystal XRD data or can be calculated computationally (**Figure 14c**). The nature of backbone, N-substituents, and the ring size are the various factors that influence the percentage of buried volume (% V_{bur}).

Several advancements in the steric parameter determination of NHCs came up with the introduction of Salerno molecular buried volume calculation (Samb*V*ca) by Cavallo et al. [54]. The online tool developed for this method utilizes the CIF file of NHC or metal-coordinated NHC as the input file to generate a two-dimensional color-coded contour map around the carbene, from which the catalytical active pockets of the complex or the sterics around the catalytic active sites can be visualized. A detailed steric map study of various types of NHC complexes was well explored by Nolan et al. [50].

4. Catalytic applications of NHC-metal complexes

In 1962, Wanzlick et al. made the early reports on the stability and reactivity of NHCs [55]. Later in 1968, the first NHC-mercury complex was isolated [56]. Despite this isolation, the studies related to the NHCs as ligands thrived only after the first isolation report by Arduengo et al. in 1991 [1]. This easy synthetic strategy facilitated the extensive studies of NHC-metal complexes and their catalytic applications. The catalytic applications of NHCs as ligands can be broadly classified into three types, and they are (a) NHCs for organocatalysis [57–59], (b) NHCs as ligands for transition metal catalysis, and (c) NHCs and main group elements-based adducts for catalysis [60]. The following section of this chapter focuses on few selective and important NHC-transition metal catalyzed reactions (**Figure 15**), since the complete details are beyond the scope of this chapter.

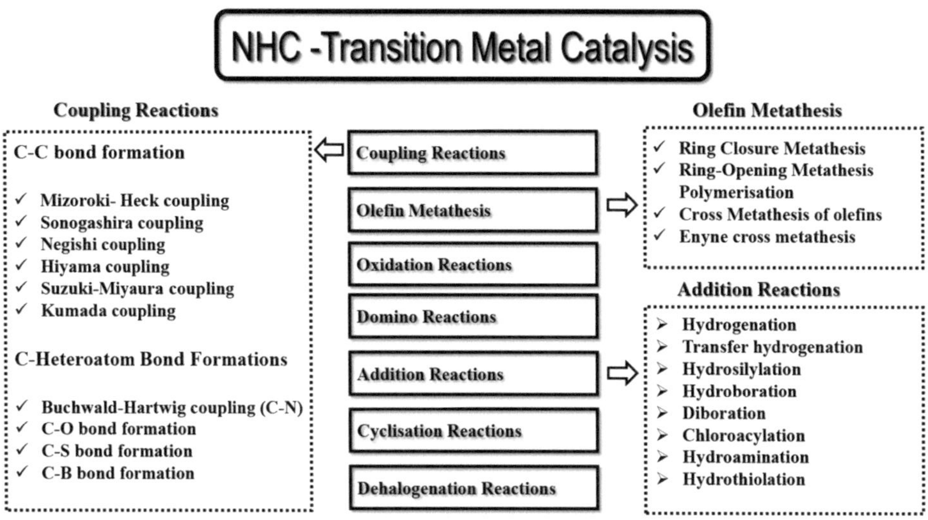

Figure 15.
Catalytic applications of NHC-transition metal complexes [14].

4.1 NHC-transition metal catalyzed cross-coupling reactions

The coupling reactions catalyzed by transition metal-NHC complexes involve both C-C bond formation reactions such as Mizoroki-Heck [61, 62], Sonogashira [63–65], Negishi [66–68], Hiyama [69–71], Suzuki-Miyaura [72–74], Kumada coupling [75, 76], C-X (X = B, N, O) bond formation reactions such as Buchwald-Hartwig amination [77–79] and Ullmann coupling [80–82] etc. Recently, the coupling reactions catalyzed by the NHC-transition metal complexes have found to be efficient with commercially viable metal salts such as iron, cobalt, and nickel. For example, the cross-coupling of methyl sulfonates and amines to afford sulfonamides uses Ni/NHC system to form the S-N bond [83]. Few well-known NHC-metal catalyzed reactions are shown in (**Figure 16**).

Recently, the Pd-NHC catalyzed metathesis of carbon-sulfur bonds seems to be one of the significant reactions, which is capable of altering the trends in the retrosynthetic approach for the preparation of organosulfur compounds [60] (**Figure 17**).

4.2 NHC-transition-metal-mediated metathesis reactions

One of the best examples to demonstrate the significance of organometallic complexes is their role as catalysts in olefin metathesis, which is having numerous synthetic applications and commercial significance. There are different types of

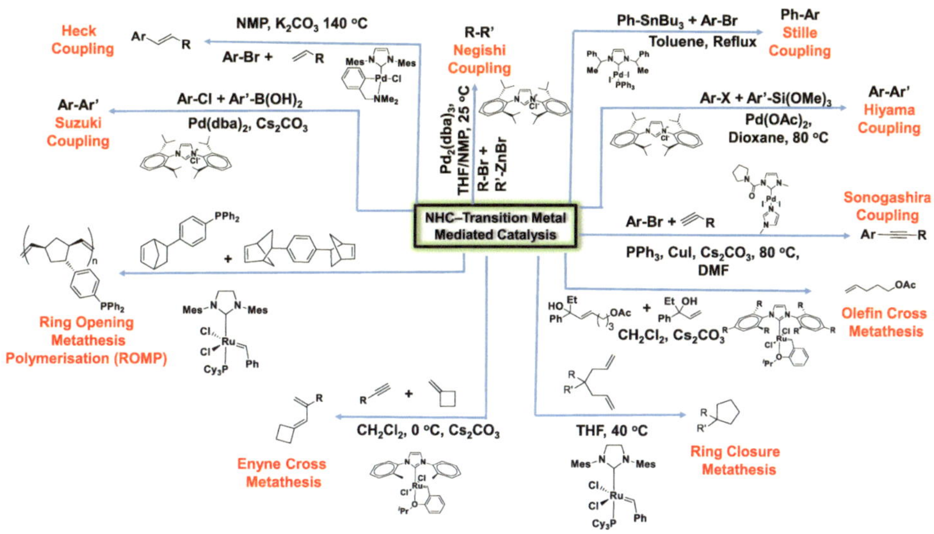

Figure 16.
Few well-known NHC-transition metal-based catalytic applications [13, 61–74].

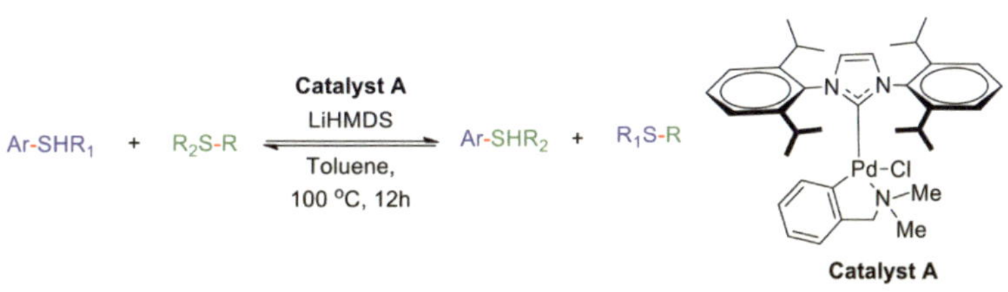

Figure 17.
Carbon-sulfur bond metathesis using Pd-NHC complex [60].

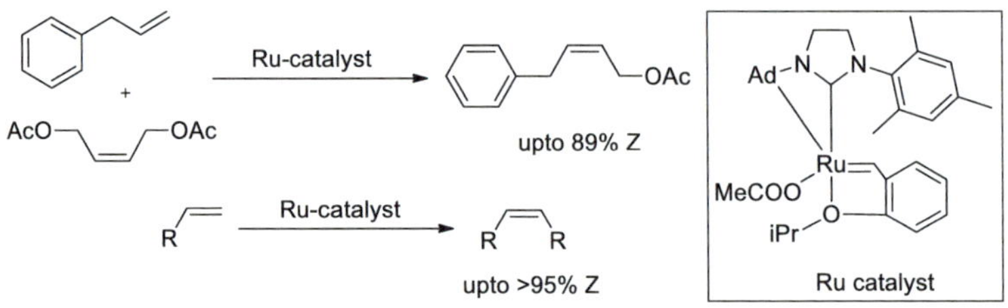

Figure 18.
NHC-Ru catalysts for Z-selective olefin cross-metathesis reaction [84].

olefin metathesis, namely ring-opening metathesis polymerization (ROMP), ring-closure metathesis (RCM), olefin cross metathesis, and enyne cross-metathesis. The ruthenium complex with a nonlabile NHC group and a labile phosphine group known to exhibit higher ring-closure metathesis (RCM) activity with better thermal stability and has tolerance toward many functional groups [14]. In general, the metathesis catalyst prefers the formation of thermodynamically favored E-isomer. However, subsequent modification on the NHC group yielded Z-isomer in significant yields. Chelating NHC-Ru complexes gave upto 95% Z selectivity. The unique steric and electronic parameters of chelating-Ru-based NHCs promote the olefin to attack cis to NHC and trans to the chelating group. This side-bound reaction mechanism results in Z-isomer as predominant product. One example of such Z-selective olefin metathesis using NHC chelated Ru-catalyst is shown (**Figure 18**) [84].

4.3 NHC-transition-metal-mediated addition reactions

Hydrogenation [85], transfer hydrogenation [86], hydrosilylation [87], hydroboration, and hydroamination [88] are the different types of NHC-transition metal catalyzed addition reactions. A brief outline of the various addition reactions catalyzed by NHC-transition metal complexes is summarized in **Figure 19**.

4.4 NHC-transition-metal-mediated catalysis—Flow chemistry

Continuous flow chemistry is a revolutionary technology that has developed rapidly during the past few years. The small size of channel reactors enables

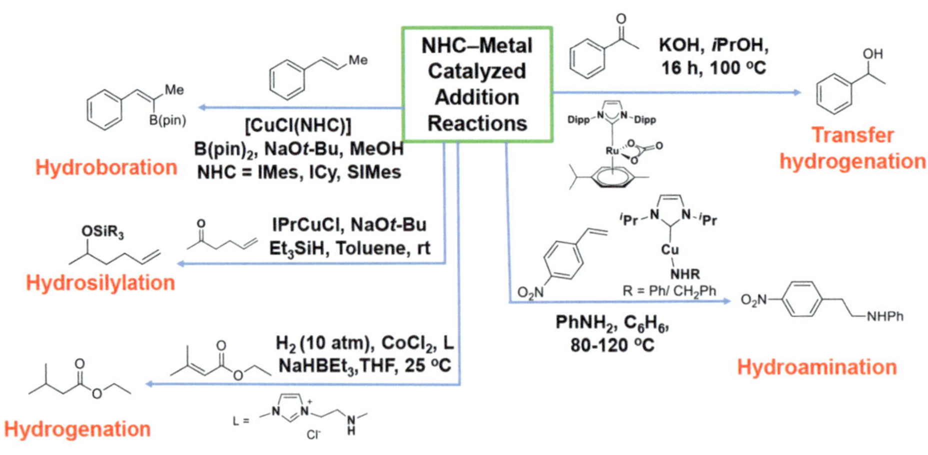

Figure 19.
Representative examples of NHC-metal catalyzed addition reactions [85–88].

increased mass transfer, efficient heat transfer, and enhanced reaction efficiency. Moreover, this method enables the handling of dangerously toxic reagents remotely and thus reduces the potential health risks [89–92]. Cole-Hamilton et al. [93] reported the first continuous flow-based olefin metathesis catalytic system with a homogeneous catalyst. In this method, the Ru-NHC catalyst was immobilized into silica pores, and CO_2 was passed as the carrier gas (**Figure 20**). It gave the metathesis products with an overall turnover number > 10,000.

Jenson et al. [94] reported a continuous nanofiltration method in which the metathesis catalyst (Ru-NHC) was allowed to react homogeneously; upon the reaction completion, the catalyst is trapped into a nano-filter setup, and it is flushed back to continue the reaction cycle. Unlike other conventional methods, there are no modifications done to the catalyst to recover and reuse it. Fabrication of NHC-metal complexes onto solid supports such as silica is another strategy used in NHC-centered flow chemistry [95–98]. Functionalized catalyst is packed into a reactor, and the reagents are forced to pass through it to yield final products (**Figure 21**).

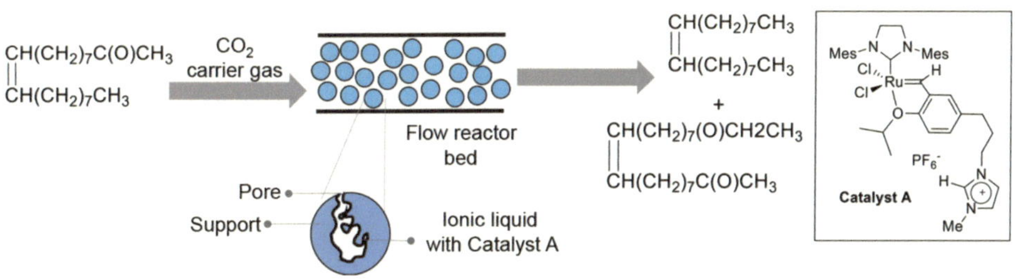

Figure 20.
Continuous flow reaction setup for olefin metathesis [93].

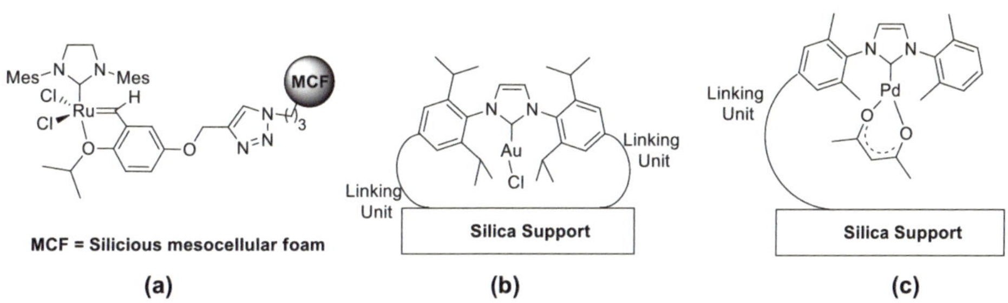

Figure 21.
Solid-supported NHC-metal catalyst and continuous flow reaction [93, 94, 99].

5. Conclusions

The imidazolium-based N-heterocyclic carbenes (NHCs) have found to be excellent candidates for improvising their properties through chemical modifications. Basic understanding and assessment of the steric and electronic factors of NHCs are crucial to explore its catalytic applications. Therefore, this chapter attempts to provide a basic glimpse of imidazolium-based NHCs through the discussions on some important synthetic routes, different methods for characterizing the electronic and steric properties of NHCs, and finally, few selective catalytic applications. Even though the literature pertaining to the imidazolium-based NHCs and its applications is vast and rapidly increasing, the discussions were limited to few important topics. Despite the extensive studies on imidazolium-based N-heterocyclic carbenes, there is plenty of scope for improvement, particularly enhancing the catalytic efficiency of NHCs with inexpensive metals, reducing the catalyst load, exploring new catalytic reactions with variety of substrates, and most significantly, making the NHC-based catalysts ready for more green and sustainable reactions.

Acknowledgements

VBG gratefully acknowledges Science and Engineering Research Board (SERB), New Delhi, for research funding (EEQ/2016/000837) and fellowship to BPJ. Thanks to the funding, it gave us an opportunity to initiate and establish our work in this exciting field on imidazolium-based NHC chemistry. Authors are also thankful to the Department of Chemistry, School of Basic and Applied Sciences, Central University of Tamil Nadu, Thiruvarur, India.

Appendices and nomenclature

Abbreviations

NHC	N-heterocyclic carbene
HOMO	highest occupied molecular orbital
LUMO	lowest unoccupied molecular orbital
TEP	Tolman electronic parameters
IAd	1,3-di(adamantyl)imidazol-2-ylidene
IMes	1,3-di(mesityl)imidazol-2-ylidene
IPr	1,3-di(2,6-diisopropyl)imidazol-2-ylidene
NMR	nuclear magnetic resonance
CV	cyclic voltammetry
SambVca	Salerno molecular buried volume calculation

Author details

Vittal B. Gudimetla*, Bony P. Joy and Sudeep Paul
Department of Chemistry, School of Basic and Applied Sciences, Central University of Tamil Nadu, Thiruvarur, Tamil Nadu, India

*Address all correspondence to: vittalbabu@cutn.ac.in

References

[1] Arduengo AJ, Harlow RL, Kline M. A stable crystalline carbene. Journal of the American Chemical Society. 1991; **113**(1):361-363. DOI: 10.1021/ja000 01a054

[2] Scholl M, Ding S, Lee CW, Grubbs RH. Synthesis and activity of a new generation of ruthenium-based olefin metathesis catalysts coordinated with 1,3-dimesityl-4,5-dihydroimidazol-2-ylidene ligands. Organic Letters. 1999; **1**(6):953-956. DOI: 10.1021/ol990909q

[3] Hopkinson MN, Richter C, Schedler M, Glorius F. An overview of N-heterocyclic carbenes. Nature. 2014; **510**(7506):485-496. DOI: 10.1038/ nature13384

[4] Herrmann WA, Köcher C. N-heterocyclic carbenes. Angewandte Chemie International Edition. 1997; **36**(20):2162-2187. DOI: 10.1002/ anie.199721621

[5] Bourissou D, Guerret O, Gabbaï FP, Bertrand G. Stable carbenes. Chemical Reviews. 2000;**100**(1):39-92. DOI: 10.1021/cr940472u

[6] Huynh HV. Electronic properties of N-heterocyclic carbenes and their experimental determination. Chemical Reviews. 2018;**118**(19):9457-9492. DOI: 10.1021/acs.chemrev.8b00067

[7] Nolan SP, editor. N-Heterocyclic Carbenes. 1st ed. Weinheim: Wiley-VCH Verlag GmbH & Co. KGaA; 2014. p. 524. DOI: 10.1002/9783527671229

[8] Huynh HV. The Organometallic Chemistry of N-Heterocyclic Carbenes. 1st ed. New Jersey: John Wiley & Sons Ltd.; 2017. p. 334. DOI: 10.1002/ 9781118698785

[9] Hu X, Tang Y, Gantzel P, Meyer K. Silver complexes of a novel tripodal N-heterocyclic carbene ligand: Evidence for significant metal-carbene π-interaction. Organometallics. 2003; **22**(4):612-614. DOI: 10.1021/om020935j

[10] Hu X, Castro-Rodriguez I, Olsen K, Meyer K. Group 11 metal complexes of N-heterocyclic carbene ligands: Nature of the metal-carbene bond. Organometallics. 2004;**23**(4):755-764. DOI: 10.1021/om0341855

[11] Scott NM, Dorta R, Stevens ED, Correa A, Cavallo L, Nolan SP. Interaction of a bulky N-heterocyclic carbene ligand with Rh(I) and Ir(I). Double C-H activation and isolation of bare 14-electron Rh(III) and Ir(III) complexes. Journal of the American Chemical Society. 2005;**127**(10): 3516-3526. DOI: 10.1021/ja043249f

[12] Cazin CSJ, editor. N-Heterocyclic Carbenes in Transition Metal Catalysis and Organocatalysis. 1st ed. Dordrecht: Springer; 2011. p. 340. DOI: 10.1007/ 978-90-481-2866-2

[13] Cavallo L, Cazin CSJ. N-heterocyclic carbenes: An introductory overview. In: Cazin CSJ, editor. N-Heterocyclic Carbenes in Transition Metal Catalysis and Organocatalysis. 1st ed. Dordrecht: Springer; 2011. pp. 1-22. DOI: 10.1007/ 978-90-481-2866-2.ch1

[14] Pugh D, Danopoulos AA. Metal complexes with 'pincer'-type ligands incorporating N-heterocyclic carbene functionalities. Coordination Chemistry Reviews. 2007;**251**(5–6):610-641. DOI: 10.1016/j.ccr.2006.08.001

[15] Benhamou L, Chardon E, Lavigne G, Bellemin-Laponnaz S, César V. Synthetic routes to N-heterocyclic carbene precursors. Chemical Reviews. 2011;**111**(4):2705-2733. DOI: 10.1021/ cr100328e

[16] Gorodetsky B, Ramnial T, Branda NR, Clyburne JAC.

Electrochemical reduction of an imidazolium cation: A convenient preparation of imidazol-2-ylidenes and their observation in an ionic liquid. Chemical Communications. 2004;**1**(17): 1972-1973. DOI: 10.1039/B407386J

[17] Aidouni A, Demonceau A, Delaude L. Microwave-assisted synthesis of N-heterocyclic carbene precursors. Synlett. 2006;**3**:493-495. DOI: 10.1055/s-2006-932455

[18] Ikhile MI, Bala MD, Nyamori VO. A greener method towards the synthesis of 1,3-diarylimidazolium tetrafluoroborates. South African Journal of Chemistry. 2011;**64**:101-104

[19] Biffis A, Baron M, Tubaro C. Poly-NHC complexes of transition metals: Recent applications and new trends. In: Pérez PJ, editor. Advances in Organometallic Chemistry. 1st ed. Vol. 63. Waltham: Elsevier Inc.; 2015. pp. 203-288. DOI: 10.1016/bs.adomc.2015.02.002.ch5

[20] De S, Udvardy A, Czégéni CE, Joó F. Poly-N-heterocyclic carbene complexes with applications in aqueous media. Coordination Chemistry Reviews. 2019; **400**:212038-212068. DOI: 10.1016/j.ccr.2019.213038

[21] Peris E, Crabtree RH. Recent homogeneous catalytic applications of chelate and pincer N-heterocyclic carbenes. Coordination Chemistry Reviews. 2004;**248**(21–24):2239-2246. DOI: 10.1016/j.ccr.2004.04.014

[22] Poyatos M, Mata JA, Peris E. Complexes with poly(N-heterocyclic carbene) ligands: Structural features and catalytic applications. Chemical Reviews. 2009;**109**(8):3677-3707. DOI: 10.1021/cr800501s

[23] Mata JA, Poyatos M, Peris E. Structural and catalytic properties of chelating bis- and tris-N-heterocyclic carbenes. Coordination Chemistry Reviews. 2007;**251**(5–6):841-859. DOI: 10.1016/j.ccr.2006.06.008

[24] Tolman CA. Steric effects of phosphorus ligands in organometallic chemistry and homogeneous catalysis. Chemical Reviews. 1977;**77**(3):313-348. DOI: 10.1021/cr60307a002

[25] Lappert MF, Pye PL. Carbene complexes. Part 12. Electron-rich olefin-derived neutral mono- and bis-(carbene) complexes of low-oxidation-state manganese, iron, cobalt, nickel, and ruthenium. Journal of the Chemical Society, Dalton Transactions. 1977;**21**:72-80. DOI: 10.1039/DT9770002172

[26] Chianese AR, Li X, Janzen MC, Faller JW, Crabtree RH. Rhodium and iridium complexes of N-heterocyclic carbenes via transmetalation: Structure and dynamics. Organometallics. 2003; **22**(8):1663-1667. DOI: 10.1021/om021029+

[27] Kelly RA, Clavier H, Giudice S, Scott NM, Stevens ED, Bordner J, et al. Determination of N-heterocyclic carbene (NHC) steric and electronic parameters using the [(NHC)Ir$(CO)_2$Cl] system. Organometallics. 2008;**27**(2):202-210. DOI: 10.1021/om701001g

[28] Huynh HV, Han Y, Jothibasu R, Yang JA. ^{13}C NMR spectroscopic determination of ligand donor strengths using N-heterocyclic carbene complexes of palladium(II). Organometallics. 2009;**28**(18):5395-5404. DOI: 10.1021/om900667d

[29] Back O, Henry-Ellinger M, Martin CD, Martin D, Bertrand G. ^{31}P NMR chemical shifts of carbene-phosphinidene adducts as an indicator of the π-accepting properties of carbenes. Angewandte Chemie International Edition. 2013;**52**(10): 2939-2943. DOI: 10.1002/anie.2012 09109

[30] Liske A, Verlinden K, Buhl H, Schaper K, Ganter C. Determining the π-acceptor properties of N-heterocyclic carbenes by measuring the ^{77}Se NMR chemical shifts of their selenium adducts. Organometallics. 2013;**32**(19): 5269-5272. DOI: 10.1021/om400858y

[31] Verlinden K, Buhl H, Frank W, Ganter C. Determining the ligand properties of N-heterocyclic carbenes from ^{77}Se NMR parameters. European Journal of Inorganic Chemistry. 2015; **2015**(14):2416-2425. DOI: 10.1002/ejic.201500174

[32] Meng G, Kakalis L, Nolan SP, Szostak M. A simple ^{1}H NMR method for determining the σ-donor properties of N-heterocyclic carbenes. Tetrahedron Letters. 2019;**60**(4):378-381. DOI: 10.1016/j.tetlet.2018.12.059

[33] Dunn MH, Konstandaras N, Cole ML, Harper JB. Targeted and systematic approach to the study of pKa values of imidazolium salts in dimethyl sulfoxide. The Journal of Organic Chemistry. 2017;**82**(14):7324-7331. DOI: 10.1021/acs.joc.7b00716

[34] Alder RW, Allen PR, Williams SJ. Stable carbenes as strong bases. Journal of the Chemical Society, Chemical Communications. 1995;**12**:1267-1268. DOI: 10.1039/C39950001267

[35] Kim YJ, Streitwieser A. Basicity of a stable carbene, 1,3-di-tert-butylimidazol-2-ylidene, in THF. Journal of the American Chemical Society. 2002;**124**(20):5757-5761. DOI: 10.1021/ja025628j

[36] Chu Y, Deng H, Cheng J, June RV. An acidity scale of 1,3-dialkylimidazolium salts in dimethyl sulfoxide solution. The Journal of Organic Chemistry. 2007;**72**(4): 7790-7793. DOI: 10.1021/jo070973i

[37] Amyes TL, Diver ST, Richard JP, Rivas FM, Toth K. Formation and stability of N-heterocyclic carbenes in water: The carbon acid pKa of imidazolium cations in aqueous solution. Journal of the American Chemical Society. 2004;**126**(13): 4366-4374. DOI: 10.1021/ja039890j

[38] Massey RS, Collett CJ, Lindsay AG, Smith AD, O'Donoghue AC. Proton transfer reactions of triazol-3-ylidenes: Kinetic acidities and carbon acid pKa values for twenty triazolium salts in aqueous solution. Journal of the American Chemical Society. 2012; **134**(50):20421-20432. DOI: 10.1021/ja308420c

[39] Enders D, Breuer K, Raabe G, Simonet J, Ghanimi A, Stegmann HB, et al. A stable carbene as π-acceptor electrochemical reduction to the radical anion. Tetrahedron Letters. 1997;**38**(16): 2833-2836. DOI: 10.1016/S0040-4039(97)00493-0

[40] Ogawa KA, Boydston AJ. Electrochemical characterization of azolium salts. Chemistry Letters. 2014; **43**(6):907-909. DOI: 10.1246/cl.140162

[41] Schotten C, Bourne RA, Kapur N, Nguyen BN, Willans CE. Electrochemical generation of N-heterocyclic carbenes for use in synthesis and catalysis. Advanced Synthesis & Catalysis. 2021;**363**(13): 3189-3200. DOI: 10.1002/adsc.202100264

[42] Feroci M, Chiarotto I, D'Anna F, Gala F, Noto R, Ornano L, et al. N-heterocyclic carbenes and parent cations: Acidity, nucleophilicity, stability, and hydrogen bonding-electrochemical study and ab initio calculations. ChemElectroChem. 2016; **3**(7):1133-1141. DOI: 10.1002/celc.201600187

[43] Feroci M, Chiarotto I, Inesi A. Advances in the knowledge of N-heterocyclic carbenes properties. The backing of the electrochemical

investigation. Catalysts. 2016;**6**(11):178. DOI: 2073-4344/6/11/178

[44] Galuppo C, Alvarenga J, Queiroz AC, Messias I, Nagao R, Abbehausen C. The electrosynthesis of gold(I) complexes: A clean, one-pot method. Electrochemistry Communications. 2020;**110**:106620. DOI: 10.1016/j.elecom.2019.106620

[45] Lake BRM, Bullough EK, Williams TJ, Whitwood AC, Little MA, Willans CE. Simple and versatile selective synthesis of neutral and cationic copper(i) N-heterocyclic carbene complexes using an electrochemical procedure. Chemical Communications. 2012;**48**(40):4887. DOI: 10.1039/C2CC30862B

[46] Chapman MR, Henkelis SE, Kapur N, Nguyen BN, Willans CE. A straightforward electrochemical approach to imine- and amine-bisphenolate metal complexes with facile control over metal oxidation state. ChemistryOpen. 2016;**5**(4):351-356. DOI: 10.1002/open.201600019

[47] Stephen HR, Schotten C, Nicholls TP, Woodward M, Bourne RA, Kapur N, et al. A versatile electrochemical batch reactor for synthetic organic and inorganic transformations and analytical electrochemistry. Organic Process Research & Development. 2020;**24**(6): 1084-1089. DOI: 10.1021/acs.oprd.0c00091

[48] Schotten C, Taylor CJ, Bourne RA, Chamberlain TW, Nguyen BN, Kapur N, et al. Alternating polarity for enhanced electrochemical synthesis. Reaction Chemistry & Engineering. 2021;**6**(1):147-151. DOI: 10.1039/D0RE00399A

[49] Fortman GC, Slawin AMZ, Nolan SP. Flexible cycloalkyl-substituted N-heterocyclic carbenes. Dalton Transactions. 2010;**39**(16): 3923-3930. DOI: 10.1039/C003214J

[50] Gómez-Suárez A, Nelson DJ, Nolan SP. Quantifying and understanding the steric properties of N-heterocyclic carbenes. Chemical Communications. 2017;**53**(18): 2650-2660. DOI: 10.1039/C7CC00255F

[51] Jover J, Cirera J. Computational assessment on the Tolman cone angles for P-ligands. Dalton Transactions. 2019;**48**(40):15036-15048. DOI: 10.1039/C9DT02876E

[52] Ferguson G, Roberts PJ, Alyea EC, Khan M. Cone angle and ligand profile calculations for bulky phosphine ligands. Inorganic Chemistry. 1978; **17**(10):2965-2967. DOI: 10.1021/ic50188a058

[53] Hillier AC, Sommer WJ, Yong BS, Petersen JL, Cavallo L, Nolan SP. A combined experimental and theoretical study examining the binding of N-heterocyclic carbenes (NHC) to the Cp*RuCl (Cp* = η^5-C_5Me_5) moiety: Insight into stereoelectronic differences between unsaturated and saturated NHC ligands. Organometallics. 2003; **22**(21):4322-4326. DOI: 10.1021/om034016k

[54] Falivene L, Cao Z, Petta A, Serra L, Poater A, Oliva R, et al. Towards the online computer-aided design of catalytic pockets. Nature Chemistry. 2019;**11**(10):872-879. DOI: 10.1038/s41557-019-0319-5

[55] Wanzlick HW. Aspects of nucleophilic carbene chemistry. Angewandte Chemie International Edition. 1962;**1**(2):75-80. DOI: 10.1002/anie.196200751

[56] Wanzlick H-W, Schönherr H-J. Direct synthesis of a mercury salt carbene complex. Angewandte Chemie International Edition. 1968;7(2): 141-142. DOI: 10.1002/anie.196801412

[57] Enders D, Niemeier O, Henseler A. Organocatalysis by N-heterocyclic carbenes. Chemical Reviews. 2007;**107**: 5606-5655. DOI: 10.1021/cr068372z

[58] Hollóczki O. The mechanism of N-heterocyclic carbene organocatalysis through a magnifying glass. Chemistry—a European Journal. 2020; **26**(22):4885-4894. DOI: 10.1002/chem. 201903021

[59] Wang N, Xu J, Lee JK. The importance of N-heterocyclic carbene basicity in organocatalysis. Organic & Biomolecular Chemistry. 2018;**16**(37): 8230-8244. DOI: 10.1039/C8OB01667D

[60] Doddi A, Peters M, Tamm M. N-heterocyclic carbene adducts of main group elements and their use as ligands in transition metal chemistry. Chemical Reviews. 2019;**119**(12):6994-7112. DOI: 10.1021/acs.chemrev.8b00791

[61] Astakhov AV, Khazipov OV, Chernenko AY, Pasyukov DV, Kashin AS, Gordeev EG, et al. A new mode of operation of Pd-NHC systems studied in a catalytic Mizoroki-Heck reaction. Organometallics. 2017;**36**(10): 1981-1992. DOI: 10.1021/acs.organo met.7b00184

[62] Balinge KR, Bhagat PR. Palladium-N-heterocyclic carbene complexes for the Mizoroki-Heck reaction: An appraisal. Comptes Rendus Chimie. 2017;**20**(7):773-804. DOI: 10.1016/ j.crci.2017.03.003

[63] Wang Z, Zheng T, Sun H, Li X, Fuhr O, Fenske D. Sonogashira reactions of alkyl halides catalyzed by NHC [CNN] pincer nickel(II) complexes. New Journal of Chemistry. 2018;**42**(14): 11465-11470. DOI: 10.1039/ C8NJ02138D

[64] Gallop CWD, Chen MT, Navarro O. Sonogashira couplings catalyzed by collaborative (N-heterocyclic carbene)-copper and palladium complexes. Organic Letters. 2014;**16**(14):3724-3727. DOI: 10.1021/ol501540w

[65] Eremin DB, Boiko DA, Kostyukovich AY, Burykina JV, Denisova EA, Anania M, et al. Mechanistic study of Pd/NHC-catalyzed sonogashira reaction: Discovery of NHC-ethynyl coupling process. Chemistry—a European Journal. 2020; **26**(67):15672-15681. DOI: 10.1002/ chem.202003533

[66] Organ MG, Avola S, Dubovyk I, Hadei N, Kantchev EAB, O'Brien CJ, et al. A user-friendly, all-purpose Pd-NHC (NHC = N-heterocyclic carbene) precatalyst for the Negishi reaction: A step towards a universal cross-coupling catalyst. Chemistry—a European Journal. 2006;**12**(18):4749-4755. DOI: 10.1002/chem.200600206

[67] Liu Z, Dong N, Xu M, Sun Z, Tu T. Mild negishi cross-coupling reactions catalyzed by acenaphthoimidazolylidene palladium complexes at low catalyst loadings. The Journal of Organic Chemistry. 2013;**78**(15):7436-7444. DOI: 10.1021/jo400803s

[68] Hadei N, Kantchev EAB, O'Brien CJ, Organ MG. The first Negishi cross-coupling reaction of two alkyl centers utilizing a Pd-N-heterocyclic carbene (NHC) catalyst. Organic Letters. 2005;7(17):3805-3807. DOI: 10.1021/ol0514909

[69] Yuen OY, So CM, Man HW, Kwong FY. A general palladium-catalyzed Hiyama cross-coupling reaction of aryl and heteroaryl chlorides. Chemistry—a European Journal. 2016; **22**(19):6471-6476. DOI: 10.1002/chem. 201600420

[70] Gu ZS, Shao LX, Lu JM. NHC-Pd(II)-Im (NHC = N-heterocyclic carbene; Im = 1-methylimidazole) complex catalyzed Hiyama reaction of aryl chlorides with aryltrimethoxysilanes. Journal of Organometallic Chemistry. 2012;**700**:

132-134. DOI: 10.1016/j.jorganchem.2011.11.030

[71] Yang J, Wang L. Synthesis and characterization of dinuclear NHC-palladium complexes and their applications in the Hiyama reactions of aryltrialkyoxysilanes with aryl chlorides. Dalton Transactions. 2012;**41**(39): 12031-12037. DOI: 10.1039/C2DT31174G

[72] Wang Q, Dai Z, Di X, Huang QF, Wang Y, Zhu J. Pd(NHC)(cinnamyl)Cl-catalyzed Suzuki cross-coupling reaction of aryl sulfonates with arylboronic acids. Molecular Diversity. 2020;**24**(4):903-911. DOI: 10.1007/s11030-019-10001-4

[73] Zhou J, Berthel JHJ, Kuntze-Fechner MW, Friedrich A, Marder TB, Radius U. NHC nickel-catalyzed Suzuki-Miyaura cross-coupling reactions of aryl boronate esters with perfluorobenzenes. The Journal of Organic Chemistry. 2016; **81**(13):5789-5794. DOI: 10.1021/acs.joc.6b01041

[74] Kaloğlu N, Özdemir İ. PEPPSI-Pd-NHC catalyzed Suzuki-Miyaura cross-coupling reactions in aqueous media. Tetrahedron. 2019;**75**(15):2306-2313. DOI: 10.1016/j.tet.2019.02.062

[75] Xi Z, Liu B, Chen W. Room-temperature Kumada cross-coupling of unactivated aryl chlorides catalyzed by N-heterocylic carbene-based nickel(II) complexes. The Journal of Organic Chemistry. 2008;**73**(10):3954-3957. DOI: 10.1021/jo800197u

[76] Ando S, Mawatari M, Matsunaga H, Ishizuka T. An N-heterocyclic carbene-based nickel catalyst for the Kumada-Tamao-Corriu coupling of aryl bromides and tertiary alkyl Grignard reagents. Tetrahedron Letters. 2016;**57**(30): 3287-3290. DOI: 10.1016/j.tetlet.2016.06.042

[77] Chartoire A, Frogneux X, Nolan SP. An efficient palladium-NHC (NHC = N-heterocyclic carbene) and aryl amination pre-catalyst: [Pd(IPr*)(cinnamyl)Cl]. Advanced Synthesis & Catalysis. 2012;**354**(10):1897-1901. DOI: 10.1002/adsc.201200207

[78] Organ MG, Abdel-Hadi M, Avola S, Dubovyk I, Hadei N, Kantchev EAB, et al. Pd-catalyzed aryl amination mediated by well defined, N-heterocyclic carbene (NHC)-Pd precatalysts, PEPPSI. Chemistry—a European Journal. 2008; **14**(8):2443-2452. DOI: 10.1002/chem.200701621

[79] Zhang Y, Lavigne G, César V. Buchwald-Hartwig amination of (hetero)aryl tosylates using a well-defined N-heterocyclic carbene/palladium(II) precatalyst. The Journal of Organic Chemistry. 2015;**80**(15): 7666-7673. DOI: 10.1021/acs.joc.5b01272

[80] Paul S, Joy BP, Sasikala G, Raghuthaman AG, Gudimetla VB. Copper-NHC based ullmann catalysis in water for selective N-arylation of 3-aminophenols. ChemistrySelect. 2020; **5**(47):15004-15009. DOI: 10.1002/slct.202003455

[81] Paul S, Joy BP, Rajendran R, Gudimetla VB. Cost efficient synthesis of diaryl ethers catalysed by CuI, imidazolium chloride and Cs_2CO_3. ChemistrySelect. 2019;**4**(24):7181-7186. DOI: 10.1002/slct.201900473

[82] Wu J-P, Saha AK, Haddad N, Busacca CA, Lorenz JC, Lee H, et al. Simple N-heterocyclic carbenes as ligands in Ullmann-type ether and thioether formations. Advanced Synthesis & Catalysis. 2016;**358**(12):1924-1928. DOI: 10.1002/adsc.201600316

[83] Li GJ, Pan YL, Liu YL, Xu HF, Chen JZ. Ni/NHC-catalyzed cross-coupling of methyl sulfinates and amines for direct access to sulfinamides. Tetrahedron Letters. 2019;**60**(46): 151260. DOI: 10.1016/j.tetlet.2019.151260

[84] Endo K, Grubbs RH. Chelated ruthenium catalysts for Z-selective olefin metathesis. Journal of the American Chemical Society. 2011; **133**(22):8525-8527. DOI: 10.1021/ja202818v

[85] Wei Z, Wang Y, Li Y, Ferraccioli R, Liu Q. Bidentate NHC-cobalt catalysts for the hydrogenation of hindered alkenes. Organometallics. 2020;**39**(17): 3082-3087. DOI: 10.1021/acs.organomet.0c00498

[86] Ma X, Guillet SG, Liu Y, Cazin CSJ, Nolan SP. Simple synthesis of [Ru(CO_3)(NHC)(p-cymene)] complexes and their use in transfer hydrogenation catalysis. Dalton Transactions. 2021; **50**(37):13012-13019. DOI: 10.1039/D1DT02098F

[87] Munro-Leighton C, Delp SA, Blue ED, Gunnoe TB. Addition of N-H and O-H bonds of amines and alcohols to electron-deficient olefins catalyzed by monomeric copper(I) systems: Reaction scope, mechanistic details, and comparison of catalyst efficiency. Organometallics. 2007;**26**(6):1483-1493. DOI: 10.1021/om061133h

[88] Kaur H, Zinn FK, Stevens ED, Nolan SP. (NHC)Cu I (NHC = N-heterocyclic carbene) complexes as efficient catalysts for the reduction of carbonyl compounds. Organometallics. 2004;**23**(5):1157-1160. DOI: 10.1021/om034285a

[89] Porta R, Benaglia M, Puglisi A. Flow chemistry: Recent developments in the synthesis of pharmaceutical products. Organic Process Research & Development. 2016;**20**(1):2-25. DOI: 10.1021/acs.oprd.5b00325

[90] Cambié D, Bottecchia C, Straathof NJW, Hessel V, Noël T. Applications of continuous-flow photochemistry in organic synthesis, material science, and water treatment. Chemical Reviews. 2016;**116**(17): 10276-10341. DOI: 10.1021/acs.chemrev.5b00707

[91] Dallinger D, Gutmann B, Kappe CO. The concept of chemical generators: On-site on-demand production of hazardous reagents in continuous flow. Accounts of Chemical Research. 2020; **53**(7):1330-1341. DOI: 10.1021/acs.accounts.0c00199

[92] Jiao J, Nie W, Yu T, Yang F, Zhang Q, Aihemaiti F, et al. Multi-step continuous-flow organic synthesis: Opportunities and challenges. Chemistry—a European Journal. 2021; **27**(15):4817-4838. DOI: 10.1002/chem.202004477

[93] Duque R, Öchsner E, Clavier H, Caijo F, Nolan SP, Mauduit M, et al. Continuous flow homogeneous alkene metathesis with built-in catalyst separation. Green Chemistry. 2011; **13**(5):1187-1195. DOI: 10.1039/C1GC15048K

[94] O'Neal EJ, Jensen KF. Continuous nanofiltration and recycle of a metathesis catalyst in a microflow system. ChemCatChem. 2014;**6**(10): 3004-3011. DOI: 0.1002/cctc.201402368

[95] Lim J, Seong Lee S, Ying JY. Mesoporous silica-supported catalysts for metathesis: Application to a circulating flow reactor. Chemical Communications. 2010;**46**(5):806-808. DOI: 10.1039/B917986K

[96] Sarmiento JT, Suárez-Pantiga S, Olmos A, Varea T, Asensio G. Silica-immobilized NHC-gold(I) complexes: Versatile catalysts for the functionalization of alkynes under batch and continuous flow conditions. ACS Catalysis. 2017;**7**(10):7146-7155. DOI: 10.1021/acscatal.7b02142

[97] Martínez A, Krinsky JL, Peñafiel I, Castillón S, Loponov K, Lapkin A, et al. Heterogenization of Pd-NHC complexes

onto a silica support and their application in Suzuki-Miyaura coupling under batch and continuous flow conditions. Catalysis Science & Technology. 2015;**5**(1):310-319. DOI: 10.1039/C4CY00829D

[98] Di Marco L, Hans M, Delaude L, Monbaliu JCM. Continuous-flow N-heterocyclic carbene generation and organocatalysis. Chemistry—a European Journal. 2016;**22**(13): 4508-4514. DOI: 10.1002/chem.201505135

[99] Autenrieth B, Frey W, Buchmeiser MR. A dicationic ruthenium alkylidene complex for continuous biphasic metathesis using monolith-supported ionic liquids. Chemistry—a European Journal. 2012;**18**(44): 14069-14078. DOI: 10.1002/chem.201201199

Chapter 8

N-Heterocyclic Carbene Mediated Organocatalysis Reactions

Yatheesh Narayana, Sandhya N.C., H.E. Dinesh, Sridhar B. Thimmaiah, Kanchugarakoppal S. Rangappa and Kempegowda Mantelingu

Abstract

Arduengo et al., isolated the first 'bottleable' carbene, the first N-heterocyclic carbene (NHC) 1,3-di(adamantyl)imidazol-2-ylidene resulted to an explosion of experimental and theoretical studies of novel NHCs being synthesized and analyzed have huge practical significance. These compounds emerged as successful ligands for coordinating transition metals, the complexes with NHC show diverse applications in the field of catalysis and organic transformation, NHC as ligand to main group elements and their properties and applications. Here this chapter provides the concise overview of N-heterocycle carbene as an organocatalyst that provides different organic transformation on to a carbonyl group. The majority of the NHC catalyzed reactions are employed in the phenomenon of reversing the electrophilic character of carbonyl carbon to nucleophilic carbon (umpolung activity) on coordination suggests benzoin, Stetter and hydroacylation reactions. Also, non-umpolung activity of bis-electrophile α,β-unsaturated acylazoliums reaction with suitable bis-nucleophiles in the organic synthesis have been studied.

Keywords: NHC carbene, Breslow intermediate, benzoin reaction, Stetter reaction, hyroacylation reaction, annulation reaction, acylazoliums, α,β-unsaturated acylazolium

1. Introduction

The first isolable carbene stabilized with adjacent phosphorous and silicon are reported by Bertrand et al. [1]. Soon after in 1991 Arduengo et al., isolated a bottleable carbene in N-heterocyclic scaffold, said to be N-heterocyclic carbene (NHC) [2]. Substituted bulky groups in NHCs adjacent to the carbene carbon provide kinetically stabilized and sterically avoids the dimerization to corresponding olefins. Remarkable stability and simple synthetic protocols revealed enormous applications of NHCs on transition metal, main group elements and as organo catalysts highlights new area of research. From the last two decades NHC organo catalysis has shepherded to extensive applications in the carbon–carbon and carbon–heteroatom bond formation. NHCs as organocatalyst involving umpolung activity of the functional group with carbonyl carbon (majority reaction employ aldehydes as substrate) acts as a transient nucleophile rather than an electrophile. The obtained nucleophile acyl anion is commonly called as "Breslow intermediate" [3]. The purpose of this chapter, to explores the NHC catalyzed transformations in organic chemistry involving in the benzoin reaction, Stetter reaction, α-β-unsaturated aldehydes in construction of heterocycles, β-functionalization of enals, hydroacylation of double bonds and triple bonds. NHCs are also known as non-umpolung mode transformations catalysis. Specially, the mode of reactivity in the generation of α-β-unsaturated acylazolium intermediates and are intercepted with various bis-nucleophiles for the enantioselective construction of various heterocyclic compounds [4–8].

2. Benzoin reaction

In 1958, Breslow proposed a mechanism in which the reaction precedes *via* an enaminol intermediate popularly known as Breslow intermediate, using thiazonium zwitterion nothing but the resonance structure of an NHC. Deprotonation of thiazonium salt (**A**) using base generates a nucleophilic thiazolidine on addition to aromatic aldehyde furnish the tetrahedral intermediate (**C**) followed by proton transfer in order to obtain enaminol (**D**). In case of benzoin condensation, nucleophilic attack of intermediate on to another equivalent of aldehyde leads to formation of 2-hydroxy ketone (**E**) with subsequent elimination of the thiazolidine (**B**) (NHC). Transformation of carbonyl compound as transient nucleophile with NHC example of umpolung reaction and Breslow intermediate can be thought of as acyl anion equivalent (**Figure 1**) [9].

2.1 Homo benzoin reaction

Homo benzoin reactions are less challenging due to chemoselectivity issue and oxidation of aldehyde to corresponding carboxylic acid that could be avoided by coupling exclusion of oxygen. An efficient benzoin reaction catalyzed by

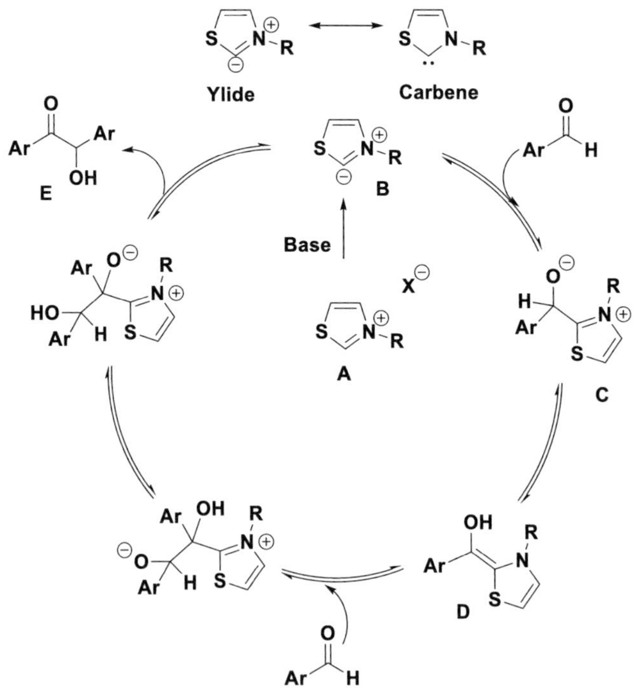

Figure 1.
Proposed mechanism of benzoin reaction.

imidazonium carbene was shown by Xu and Xia in 2005 [10] (**Figure 2**). Iwamoto et al. shows NHC catalyzed benzoin reaction in aqueous media [11, 12]. Enantioselective benzoin reaction employing NHC catalyst have many illustrations by various groups [13] (**Figure 3**). The most efficient enantioselective benzoin reaction was reported by Connon et al. with >99% ee (**Figure 4**) [14].

2.2 Cross benzoin reaction

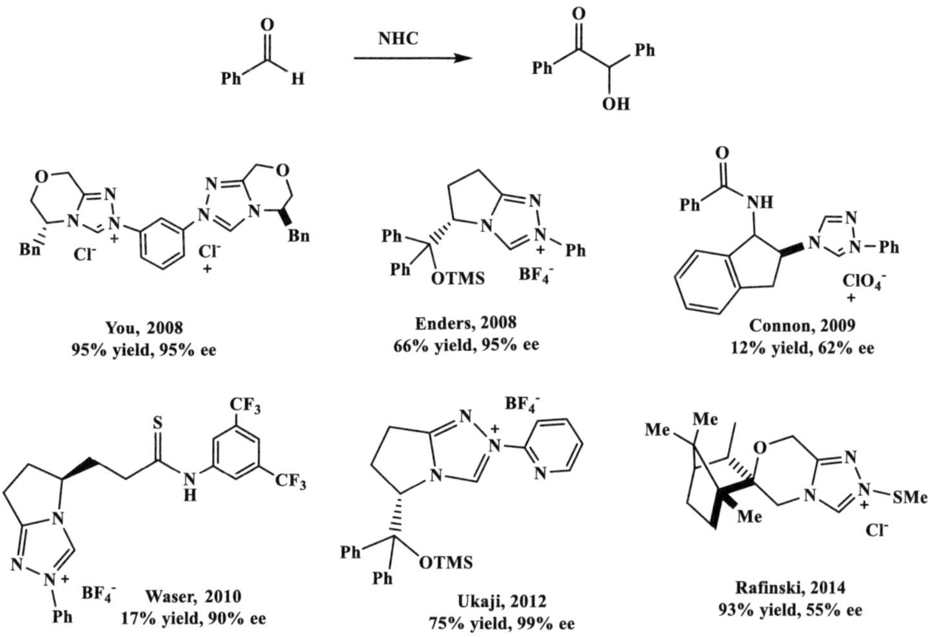

Figure 2.
Efficient enantioselective homo benzoin reaction.

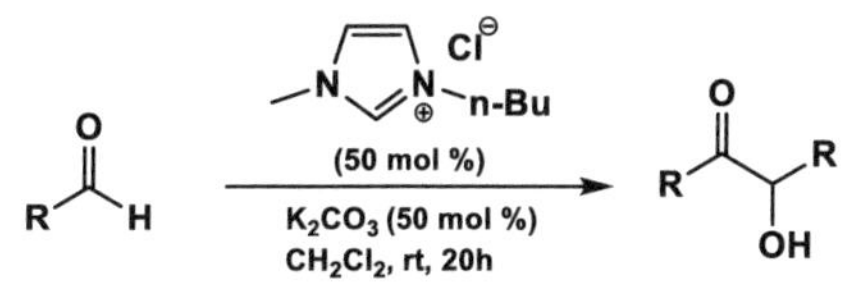

Figure 3.
An efficient benzoin reaction catalyzed by imidazonium carbene.

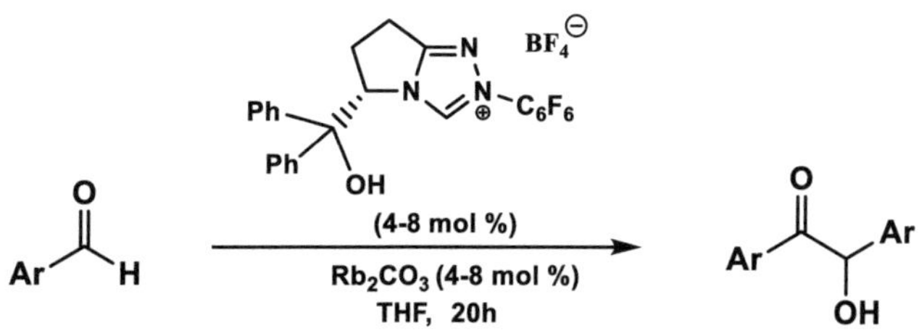

Figure 4.
The most efficient enantioselective benzoin reaction by Connon et al. with >99%.

In an NHC driven cross benzoin reaction, total four products are possible, a pair of homo benzoin and cross benzoin adducts each. Electronic or steric reasons one of the aldehydes may be significantly less reactive. Connon et al. found that cross benzoin reaction could be biased by synthetically useful amount for the synthesis of desirable cross coupling products (**Figure 5**) [15].

Glorious et al., demonstrated a highly selective cross benzoin reaction with broad substrate scope [16]. Young et al. showed reactivity controlled by the carbine catalyst. An intermolecular cross coupling of aromatic aldehyde for the formation of Breslow intermediate is followed by coupling with acetaldehyde using thiazonium carbene catalyst. In constant acyl anion generation from acetaldehyde preferred by triazolium carbine is followed by coupling with aromatic aldehyde (**Figure 6**) [17].

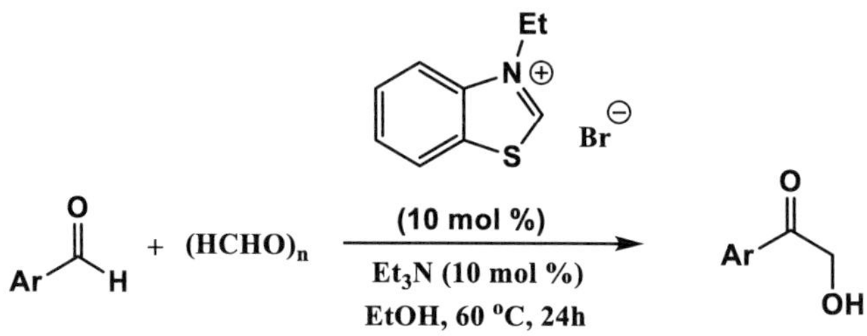

Figure 5.
Cross benzoin reaction by Connon et al.

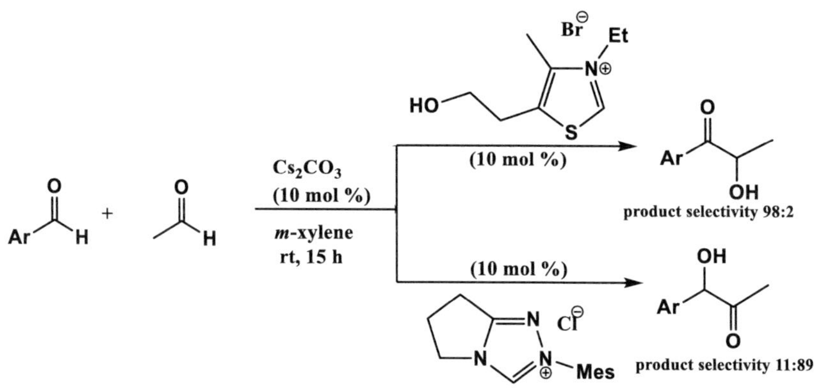

Figure 6.
Reactivity controlled reaction by using deferent carbene catalyst.

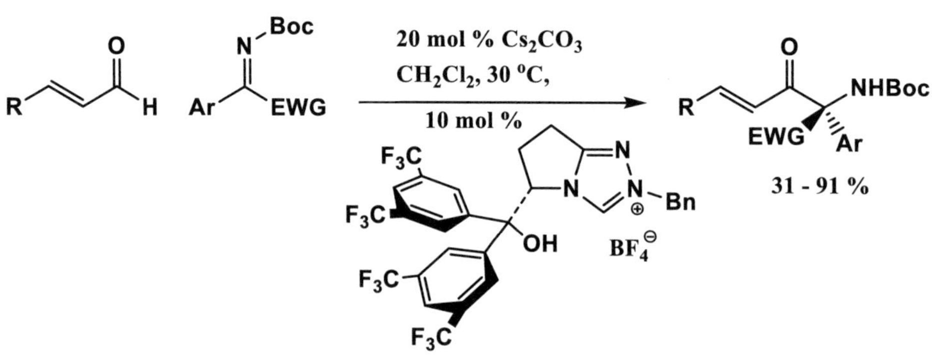

Figure 7.
The first aza benzoin reaction by Enders et al.

2.3 Aza benzoin reaction

The coupling of Breslow intermediate to imines was reported by Murry et al. for the first time [18]. NHC addition to highly electrophilic N-BOC imines leads to the formation of corresponding aza-Breslow intermediate, but this can be reversible under suitable reaction condition to obtain a pure 2-amino ketone [19]. Enders et al., reported the first aza benzoin reaction with trifluoromethyl ketimines using achiral thiazonium precatalyst [20]. Where enals reacts with trifluoromethyl ketone, obtained excellent enantioselectivity using chiral triazolium catalyst, which also furnishes electronic diminished groups tolerance (**Figure** 7) [21].

3. The Stetter reaction

In 1976, Stetter developed the thiazolium catalyzed highly selective conjugate addition reaction of aromatic/aliphatic aldehydes with an array of Michael acceptors and in most cases these reactions proceed in an intermolecular fashion [22, 23]. The Stetter reaction can be catalyzed by broad range of thiazolium, triazolium and imidazolium carbene, mostly α,β-unsaturated ketones are used as Michael acceptor. The formation of 1,4-diketone, γ-ketonitriles and γ-ketoesters resulting from NHC catalyzed Stetter reaction which is not easy by conventional method [24–27]. A mechanism of benzoin reaction of similar fashion is observed in the Stetter reaction, the in-situ generated free carbene (**B**) of azolium salt (**A**) when treated with base reacts with aldehyde, generating the nucleophilic Breslow intermediate (**D**). This intermediate undergoes irreversible addition to the Michael acceptor generating intermediate (**F**), which on proton transfer and subsequent release of free carbene affords the desired Stetter product (**G**) (**Figure 8**) [28].

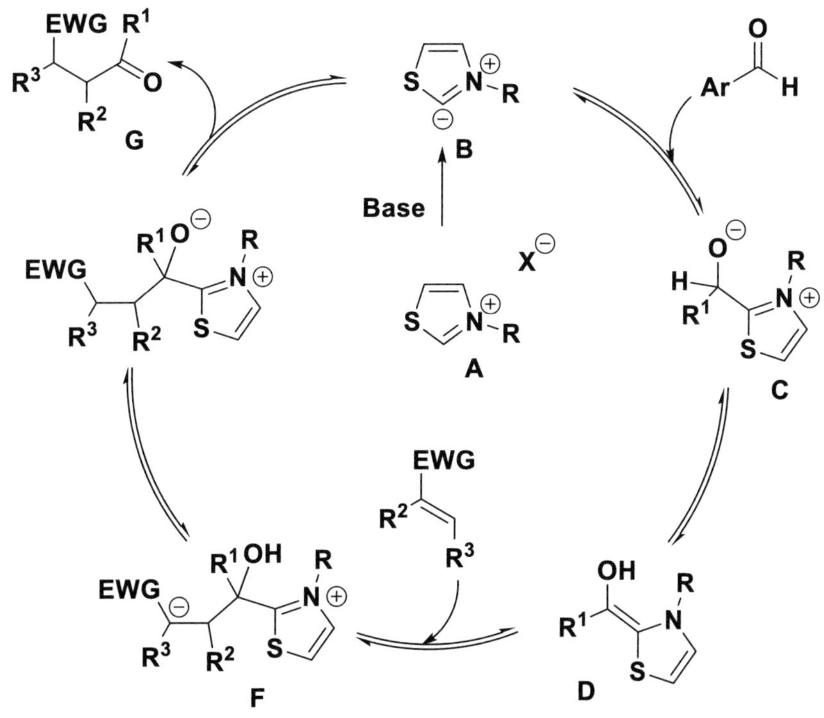

Figure 8.
Proposed mechanism of Stetter reaction.

3.1 Intramolecular Stetter reaction

The first general intra molecular Stetter reaction was reported by Ciganek in 1995. By the next year, Enders et al. explains the first enantioselective intramolecular Stetter reaction. Later, the implementation of chiral NHC's resulted in asymmetric transformation leading to synthesis of enantioselective 1,4-bifunctional compounds. The amino indanol derived chiral triazolium salt and pentafluoroamyl-substituted were developed by Rovis et al. in 2002 for the most efficient catalyst for the enantioselective intramolecular reaction (**Figure 9**) [29–33].

3.2 Intermolecular Stetter reaction

The initial intermolecular reaction established by Endres et al. with chiral thiazolium catalyzed reaction of *n*-butanal with chalcones resulted in Stetter product with 40% ee [34, 35] (**Figure 10**).

Rovis et al. contributed in the asymmetric intermolecular Stetter reaction of glyoxamide derivatives as aldehyde component and alkylidene molecule as Michael acceptors [36, 37]. DiRocco and Rovis expanded the reaction of intermolecular

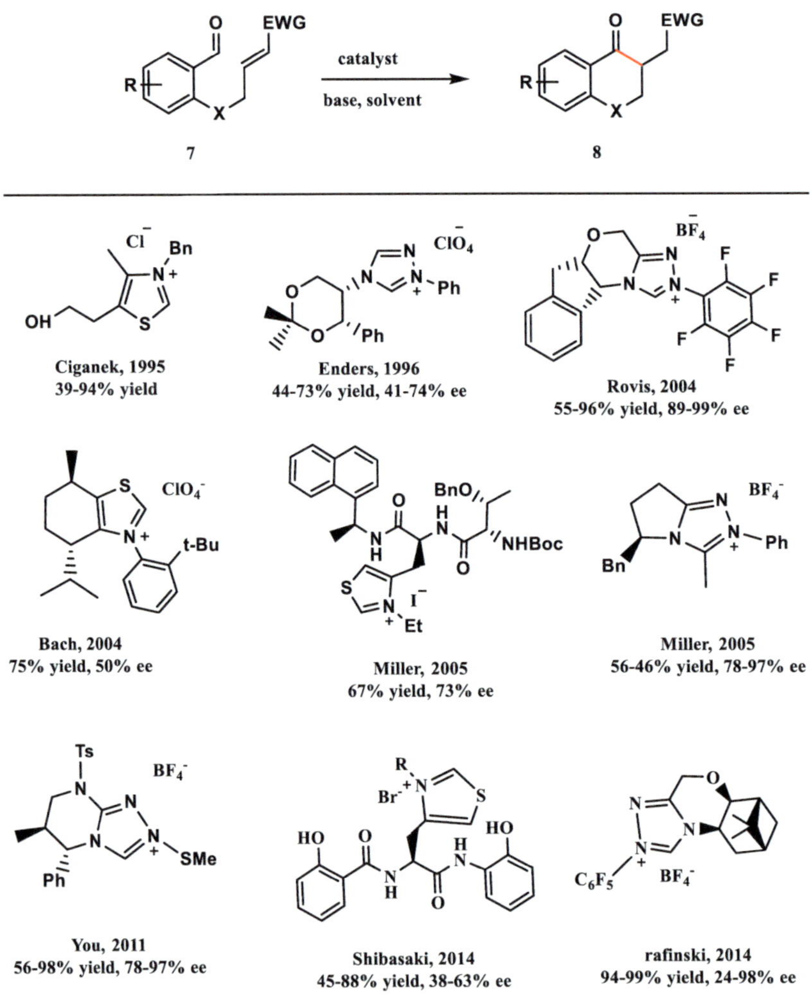

Figure 9.
Some examples of intramolecular Stetter reaction.

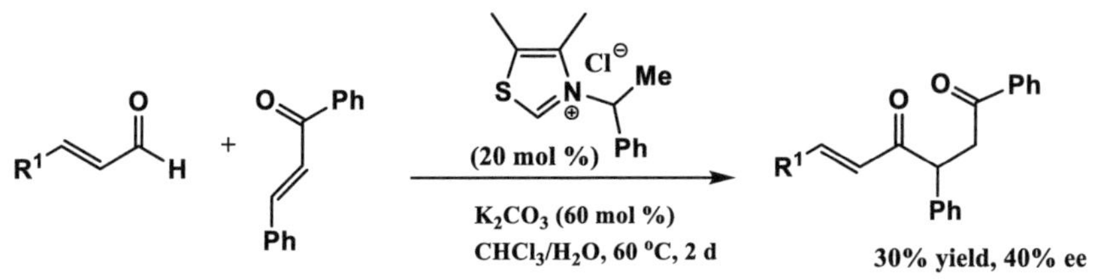

Figure 10.
The first intermolecular Stetter reaction by Enders et al.

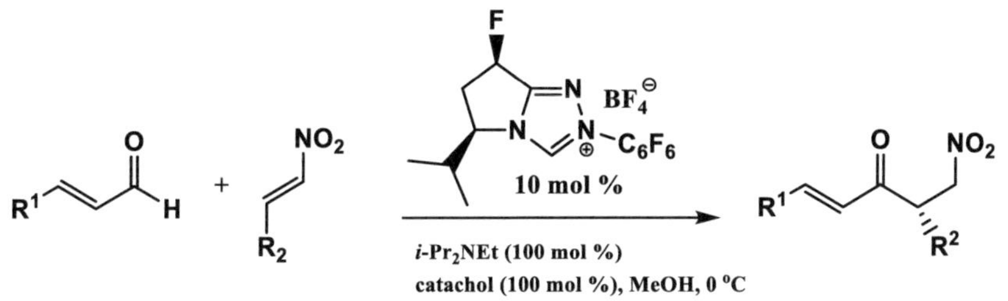

Figure 11.
Intermolecular Stetter reaction with α,β-unsaturated aldehyde and β-nitrostyrenes.

Stetter reaction with α,β-unsaturated aldehyde as aldehyde component and β-nitrostyrenes using fluorinatedtriazolium salt (**Figure 11**) [38].

The synthesis of enantiomeric α-amino acid derivative was developed by Glorius et al. using N-alylamidoarylate as Michael accepter [39]. Biju co-workers reported the efficient enantioselective NHC catalyzed intermolecular Stetter reaction of aldehydes with α,β-unsaturated sulfones and vinyl phosphonates [40, 41].

3.3 Hydroacylation of enol ethers of double bonds and triple bonds

Acyl anion reaction has been extended followed by the addition to electron neutral as carbon–carbon multiple bonds. She and Pan explained for the first time using alkyl tosylates (**Figure 12**) and found the selectivity changes when the native substrate with a phenyl group [42]. The reaction resulted in the formation of benzofuranone and the mechanism involving the addition of the Breslow intermediate to the C–C double bond of the enol ether [43].

A very important reaction for the synthesis of chromanones from the intermolecular cyclization of 2-allyloxy benzaldehydes using thiazolium NHC-catalyst by Glorious et al. for the first time [44] (**Figure 13**). Biju et al. showed the coupling of cyclopropenes using achiral triazolium with aryl aldehyde coupling partners (**Figure 14**) [45].

3.4 Annulation reaction

The extended Breslow intermediate has explored very well especially for the synthesis of heterocyclic compounds. Bode and Glorius reported NHC catalyzed homoenolate reactivity leading to synthesis of useful molecule from simple enal cascade. The reaction involved in the formation of extended Breslow intermediate from enal with carbene followed by 1,2-addition to aryl aldehyde which was then cyclized to deliver the γ-lactone product (**Figure 15**).

Several groups explored NHC catalyzed to synthesize enantioselective γ-lactone [46–48], spirocyclic γ-lactones [49, 50]. [3 + 4] Annulation reaction between enals

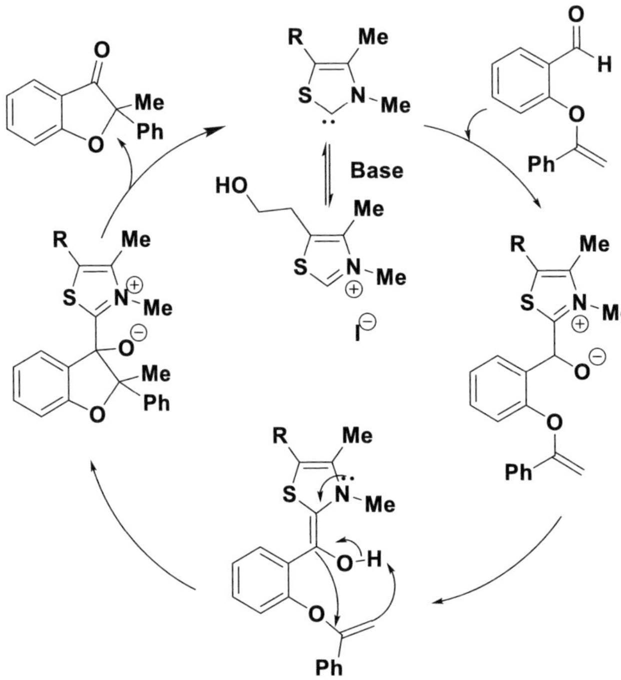

Figure 12.
Proposed mechanism of hydroacylation reaction of enol ether.

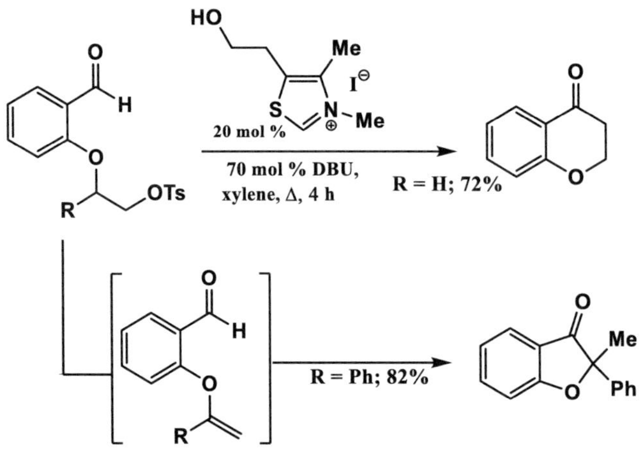

Scheme 13

Figure 13.
The intermolecular cyclization of 2-allyloxy benzaldehydes using thiazolium NHC-catalyst.

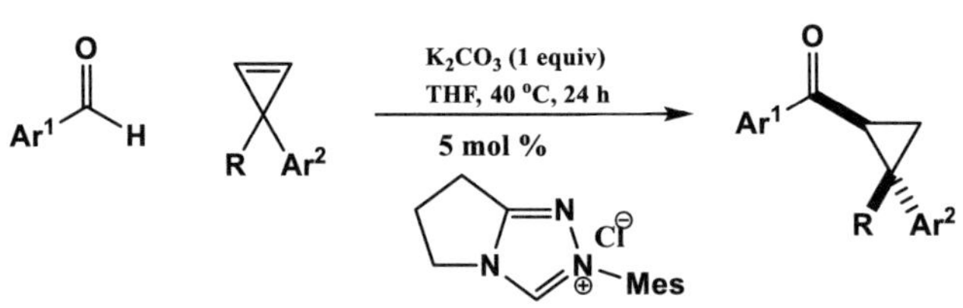

Figure 14.
The coupling of cyclopropenes using achiral triazolium with aryl aldehyde.

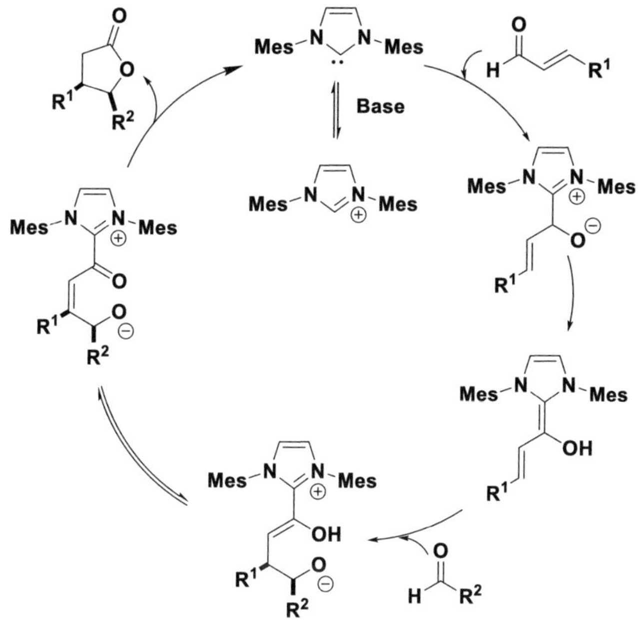

Figure 15.
Mechanism of homoenolate reactivity for the formation of γ-lactone product.

and o-quinonemethides (**Figure 16**) was reported by Ye et al. in 2013 to obtain dioxolane fused-quinone methides [51].

The [8 + 3] annulation of enals and tropone *via* conjugate addition followed by cyclization afforded a lactone (**Figure 17**), reported by Nair et al. [52].

A number of nitrogen-containing heterocycles generated by homoenolate has explored largely. Synthesis of enantioselective cyclic sulfonyl ketamine annulation [53], β-lactam formation [54], synthesis of pyrazolidinones [55], isoxazolidinone formation [56] and mainly nitroso coupling reactions [57] and others have many

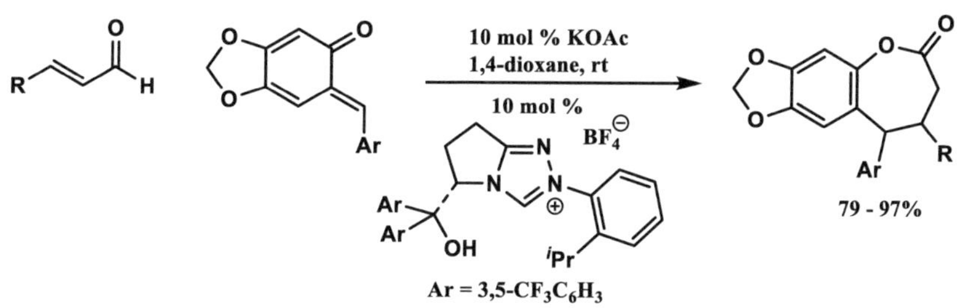

Figure 16.
[3 + 4] Annulation reaction between enals and o-quinonemethides.

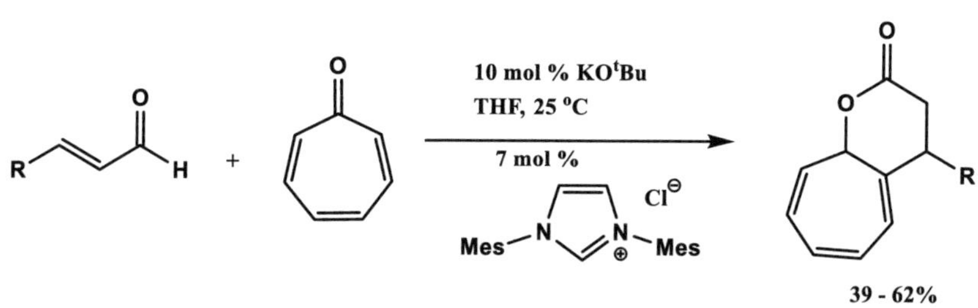

Figure 17.
[8 + 3] Annulation of enals and tropone.

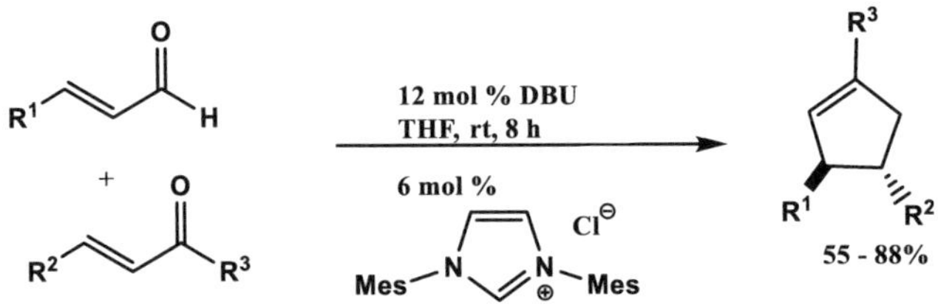

Figure 18.
The first NHC generated homoenolate in the formation of cyclopentenes.

synthetic applications. The first NHC generated homoenolate further utilized in the formation of cyclopentenes was introduced by Nair et al. in 2006 [58, 59]. The homoenolate intermediate reacts with chalcones generates allyloxide, further cyclized 4-membered β-lactone which decarboxylate to provide cyclopentene (**Figure 18**).

4. α,β-Unsaturated acylazolium intermediate

An important strategy of non-umpolung transformation reaction proceeds through the α,β-unsaturated acylazolium conjugation addition of various bis-nucleophile, a wide variety of carbo cycles and heterocycles are synthesized. 1,2-addition followed by cyclization [4, 5, 60]. α,β-unsaturated acylazoliums can be generated from α,β-unsaturated aldehyde with external oxidants [61–66], ynals, 2-bromo enals [67–69], α,β-unsaturated esters [70] or acyl fluorides [71, 72], etc. (**Figure 19**).

The main application of acylazolium intermediate found in the biosynthesis of clavulanic acid (potent β-lactamase inhibitor) from conjugation addition of L-arginine to the α,β-unsaturated acylazolium, demonstrated by Merski and Townsend (**Figure 20**) [73, 74].

Lupton reported the Claisen type reaction of α,β-unsaturated enol esters to the corresponding acylazolium/enolate pair followed by rearranged to 2,3-dihydropyrozones (**Figure 21**) [71, 72].

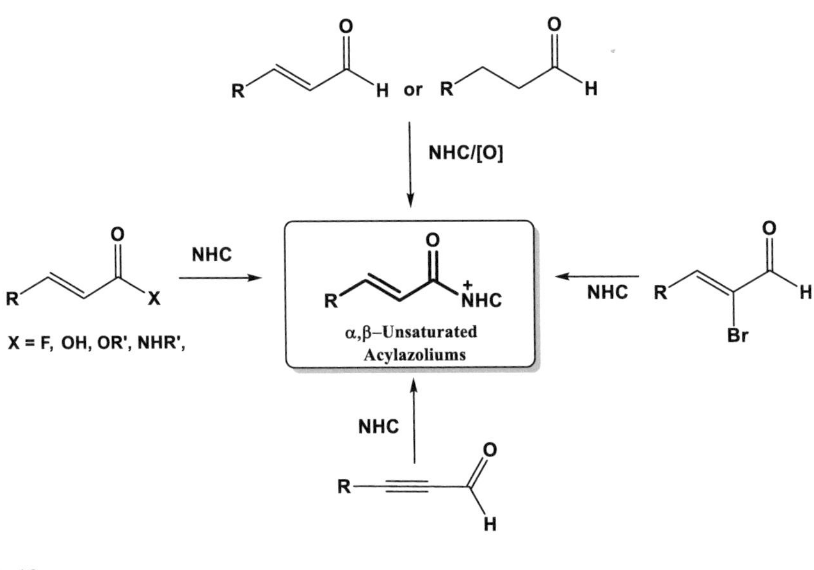

Figure 19.
Generation of α,β-unsaturated acylazoliums.

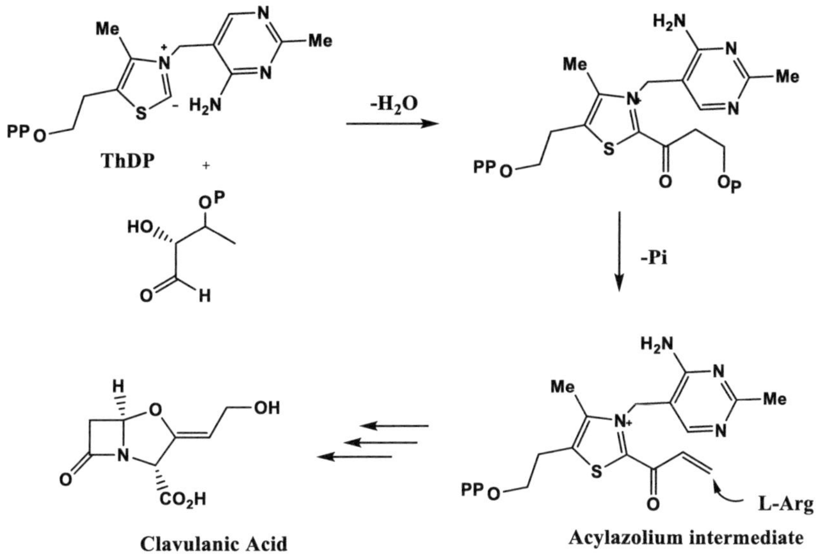

Figure 20.
Acylazolium intermediate found in the biosynthesis of clavulanic acid.

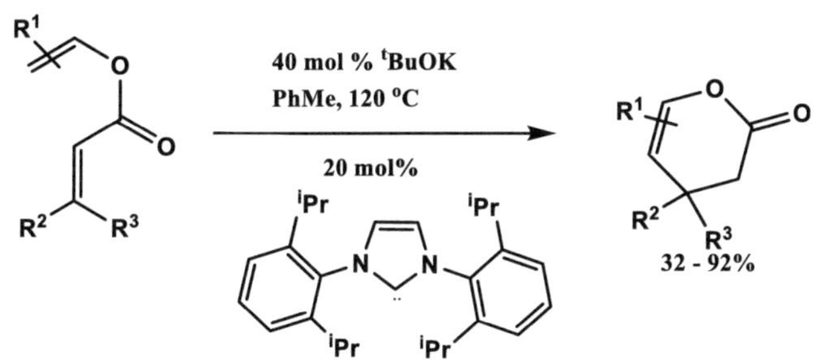

Figure 21.
Claisen type reaction of α,β-unsaturated enol esters to 2,3-dihydropyronones.

The generated electrophilic acylazolium intermediates involved in various annulation and cycloaddition reaction with bis-nucleophiles. α,β-unsaturated acylazolium with various cyclic and acyclic bis-nucleophiles for the synthesis of dihydropyranones and dihydropyridines in a formal [3 + 3] annulation reaction (**Figure 22**), demonstrated by Biju et al. [68].

Cycloaddition reaction of α-bromoenals with 1,3-diketones by using different achiral NHC catalyst obtained the product with the same absolute configuration but different stereodirecting substituents (**Figure 23**) [67].

5. Conclusion

N-heterocyclic carbenes have had a broad scope in the field of organic chemistry, often tolerating the construction of complex molecules from simple starting materials. The various modes of NHC catalysis such as the generation of Breslow intermediates, homoenolates, α,β-unsaturated acylazoliums, NHC enolates can be engaged in the synthesis of various heterocycles and carbocycles. The use of enantiomerically pure carbene catalyst can result in asymmetric synthesis of the

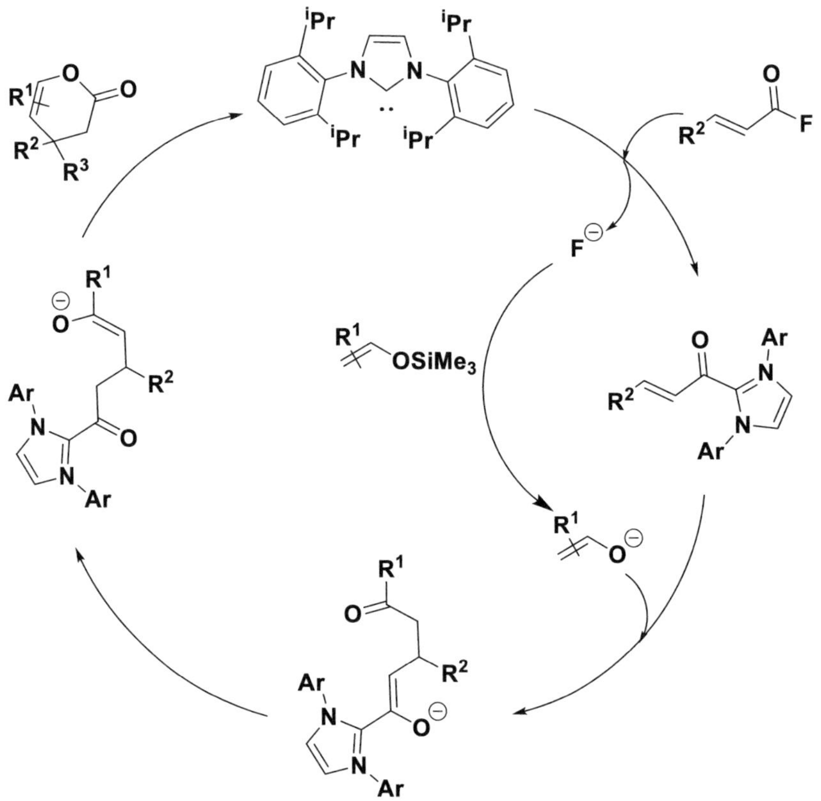

Figure 22.
The synthesis of dihydropyranones in a formal [3 + 3] annulation reaction.

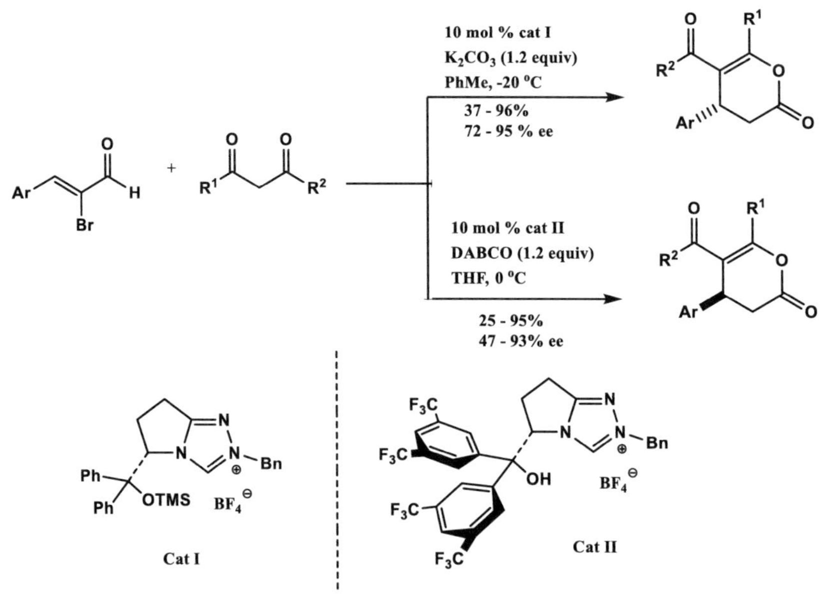

Figure 23.
Cycloaddition reaction of α-bromoenals with 1,3-diketones.

target molecules. This chapter has focused on the reactivity pathways which expand the variety reactions with suitable reaction companions beyond the old-fashioned aldehydes with enhanced catalytic conventions.

Conflict of interest

The authors declare no conflict of interest.

Author details

Yatheesh Narayana[1], Sandhya N.C.[2], H.E. Dinesh[1], Sridhar B. Thimmaiah[3], Kanchugarakoppal S. Rangappa[4*] and Kempegowda Mantelingu[1*]

1 Department of Studies in Chemistry, University of Mysore, Mysore, Karnataka, India

2 PG Department of Chemistry, St. Philomena's College, Mysore, Karnataka, India

3 Department of Studies in Chemistry, Maharani's Science College for Women, Bangalore, Karnataka, India

4 Vignana Bhavan, Institute of Excellence (IOE), University of Mysore, Mysore, Karnataka, India

*Address all correspondence to: rangappaks@gmail.com and kmlingu@gmail.com

References

[1] Aigau A, Grutzmacher H, Baceiredo A, Bertrand G. Analogous α,α'-biscarbenoid triply bonded species: Synthesis of a stable λ3-phosphinocarbene-λ5-phosphaacetylene. Journal of the American Chemical Society. 1988;**110**: 6463-6466. DOI: 10.1021/ja00227a028

[2] Arduengo AJ III, Harlow RL, Kline M. A stable crystalline carbene. Journal of the American Chemical Society. 1991;**113**:361-363. DOI: 10.1021/ja00001a054

[3] Breslow R. On the mechanism of thiamine action. IV. Evidence from studies on model systems. Journal of the American Chemical Society. 1958; **80**(14):3719-3726. DOI: 10.1021/JA01547A064

[4] Zhang C, Hooper JF, Lupton DW. N-Heterocyclic carbene catalysis via the α,β-unsaturated acyl azolium. ACS Catalysis. 2017;**7**:2583-2596. DOI: 10.1021/acscatal.6b03663

[5] Mahatthananchai J, Bode JW. On the mechanism of N-heterocyclic carbene-catalyzed reactions involving acyl azoliums. Accounts of Chemical Research. 2014;**47**:696-707. DOI: 10.1021/ar400239v

[6] Ryan SJ, Candish L, Lupton DW. Acyl anion free N-heterocyclic carbene organo catalysis. Chemical Society Reviews. 2013;**42**:4906-4917. DOI: 10.1039/C3CS35522E

[7] De Sarkar S, Biswas A, Samanta RC, Studer A. Catalysis with N-heterocyclic carbenes under oxidative conditions. Chemistry - A European Journal. 2013; **19**:4664-4678. DOI: 10.1002/chem.201203707

[8] Knappke CEI, Imami A, Jacobi von Wangelin A. Oxidative N-heterocyclic carbene catalysis. ChemCatChem. 2012; **4**:937-941. DOI: 10.1002/cctc.201200133

[9] Bugaut X, Glorius F. Organocatalytic umpolung: N-heterocyclic carbenes and beyond. Chemical Society Reviews. 2012;**41**:3511-3522. DOI: 10.1039/C2CS15333E

[10] Xu L-W, Gao Y, Yin J-J, Li L, Xia C-G. Efficient and mild benzoin condensation reaction catalyzed by simple 1-N-alkyl-3-methylimidazolium salts. Tetrahedron Letters. 2005;**46**:5317. DOI: 10.1016/j.tetlet.2005.06.015

[11] Iwamoto K, Hamaya M, Hashimoto N, Kimura H, Suzuki Y, Sato M. Benzoin reaction in water as an aqueous medium catalyzed by benzimidazolium salt. Tetrahedron Letters. 2006;**47**:7175-7177. DOI: 10.1016/j.tetlet.2006.07.153

[12] Iwamoto K, Kimura H, Oike M, Sato M. Methylene-bridged bis (benzimidazolium) salt as a highly efficient catalyst for the benzoin reaction in aqueous media. Organic & Biomolecular Chemistry. 2008;**6**: 912-915. DOI: 10.1039/B719430G

[13] Bugaut X. Benzoin and aza-benzoin. In: Molander GA, Knochel P, editors. Comprehensive Organic Synthesis. 2nd ed. Vol. 1. Oxford, UK: Elsevier; 2014. pp. 424-470

[14] Baragwanath L, Rose CA, Zeitler K, Connon SJ. Highly Enantioselective benzoin condensation reactions involving a bifunctional protic pentafluorophenyl-substituted triazolium precatalyst. The Journal of Organic Chemistry. 2009;**74**: 9214-9217. DOI: 10.1021/jo902018j

[15] Rose CA, Gundala S, Connon SJ, Zeitler K. Chemoselective crossed acyloin condensations: Catalyst and substrate control. Synthesis. 2011;**2**: 190-198. DOI: 10.1055/s-0030-1258363

[16] Kuhl N, Glorius F. Direct and efficient N-heterocyclic

carbene-catalyzed hydroxymethylation of aldehydes. Chemical Communications. 2011;**47**:573. DOI: 10.1039/C0CC02416C

[17] Jin MY, Kim SM, Han H, Ryu DH, Yang JW. Switching regioselectivity in crossed acyloin condensations between aromatic aldehydes and acetaldehyde by altering N-heterocyclic carbene catalysts. Organic Letters. 2011;**13**: 880-883. DOI: 10.1021/ol102937w

[18] Murry JA, Frantz DE, Soheili A, Tillyer R, Grabowski EJJ, Reider PJ. Synthesis of α-amido ketones via organic catalysis: Thiazolium-catalyzed cross-coupling of aldehydes with acylimines. Journal of the American Chemical Society. 2001;**123**:9696-9697. DOI: 10.1021/ja0165943

[19] DiRocco DA, Rovis T. Catalytic asymmetric cross-aza-benzoin reactions of aliphatic aldehydes with N-Boc-protected imines. Angewandte Chemie, International Edition. 2012;**51**: 5904-5906. DOI: 10.1002/anie.201202442

[20] Enders D, Henseler A, Lowins S. N-Heterocyclic carbene catalyzed nucleophilic acylation of trifluoromethyl ketimines. Synthesis. 2009;**24**:4125-4128. DOI: 10.1055/s-0029-1217070

[21] Sun L-H, Liang Z-Q, Jia W-Q, Ye S. Enantioselective N heterocyclic carbene catalyzed aza-benzoin reaction of enals with activated ketimines. Angewandte Chemie, International Edition. 2013;**52**: 5803-5806. DOI: 10.1002/ange.201301304

[22] Stetter H. Catalyzed addition of aldehydes to activated double bonds—A new synthetic approach. Angewandte Chemie (International Ed. in English). 1976;**15**:639-647. DOI: 10.1002/anie.197606391

[23] Stetter H, Kuhlmann H. In: Paquette LA, editor. Organic Reactions. Vol. 40. New York: Wiley & Sons; 1991. p. 407

[24] Read de Alaniz J, Rovis T. The catalytic asymmetric intramolecular Stetter reaction. Synlett. 2009(**8**): 1189-1207. DOI: 10.1055/s-029-1216654

[25] Rovis T. Development of chiral bicyclic triazolium salt organic catalysts: The importance of the N-aryl substituent. Chemistry Letters. 2008;**37**: 2. DOI: 10.1246/cl.2008.2

[26] Christmann M. New developments in the asymmetric Stetter reaction. Angewandte Chemie, International Edition. 2005;**44**:2632. DOI: 10.1002/anie.200500761

[27] Read de Alaniz J, Kerr MS, Moore JL, Rovis R. Scope of the asymmetric intramolecular Stetter reaction catalyzed by chiral nucleophilic triazolinylidene carbenes. The Journal of Organic Chemistry. 2008;**73**:2033. DOI: 10.1021/jo702313f

[28] Moore JL, Silvestri AP, Read de Alaniz J, DiRocco DA, Rovis T. Mechanistic Investigation of the enantioselective intramolecular Stetter reaction: Proton transfer is the first irreversible step. Organic Letters. 2011; **13**:1742-1745. DOI: 10.1021/ol200256a

[29] Kerr MS, Read de Alaniz J, Rovis T. A highly enantioselective catalytic intramolecular Stetter reaction. Journal of the American Chemical Society. 2002;**124**:10298-10299. DOI: 10.1021/ja027411v

[30] Kerr MS, Rovis T. Enantioselective synthesis of quaternary stereocenters via a catalytic asymmetric Stetter reaction. Journal of the American Chemical Society. 2004;**126**:8876-8877. DOI: 10.1021/ja047644h

[31] Kerr MS, Read de Alaniz J, Rovis T. An efficient synthesis of achiral and chiral 1,2,4-triazolium salts: Bench

stable precursors for N-heterocyclic carbenes. The Journal of Organic Chemistry. 2005;**70**:5725-5728. DOI: 10.1021/jo050645n

[32] Moore JL, Kerr MS, Rovis T. Enantioselective formation of quaternary stereocenters using the catalytic intramolecular Stetter reaction. Tetrahedron. 2006;**62**:11477-11482. DOI: 10.1016/j.tet.2006.06.042

[33] Cullen SC, Rovis T. Catalytic asymmetric Stetter reaction onto vinylphosphine oxides and vinylphosphonates. Organic Letters. 2008;**10**:3141-3144. DOI: 10.1021/ol801047k

[34] Enders D. Stereoselective Synthesis. Heidelberg: Springer-Verlag; 1993. pp. 63-90

[35] Enders D, Bockstiegel B, Dyker H, Jegelka U, Kipphardt H, Kownatka D, et al. Dechema-Monographies. Vol. 129. Weinheim: VCH; 1993. p. 209

[36] Liu Q, Rovis T. Enantio and diastereoselective intermolecular Stetter reaction of glyoxamide and alkylidene ketoamides. Organic Letters. 2009;**11**: 2856. DOI: 10.1021/ol901081a

[37] Liu Q, Perreault S, Rovis T. Catalytic asymmetric intermolecular Stetter reaction of glyoxamides with alkylidenemalonates. Journal of the American Chemical Society. 2008;**130**: 14066. DOI: 10.1021/ja805680z

[38] DiRocco DA, Rovis T. Catalytic asymmetric intermolecular Stetter reaction of enals with nitroalkenes: Enhancement of catalytic efficiency through bifunctional additives. Journal of the American Chemical Society. 2011; **133**:10402. DOI: 10.1021/ja203810b

[39] Jousseaume T, Wurz NE, Glorius F. Highly enantioselective synthesis of α-amino acid derivatives by an NHC-catalyzed intermolecular Stetter reaction. Angewandte Chemie, International Edition. 2011;**50**:1410. DOI: 10.1002/anie.201006548

[40] Bhunia A, Yetra SR, Bhojgude SS, Biju AT. Efficient synthesis of γ-keto sulfones by NHC-catalyzed intermolecular Stetter reaction. Organic Letters. 2012;**14**:2830. DOI: 10.1021/ol301045x

[41] Patra A, Bhunia A, Biju AT. Facile synthesis of γ-ketophosphonates by an intermolecular Stetter reaction onto vinylphosphonates. Organic Letters. 2014;**16**:4798. DOI: 10.1021/ol502262d

[42] He J, Zheng J, Liu X, She X, Pan X. N-Heterocyclic carbene catalyzed nucleophilic substitution reaction for construction of benzopyrones and benzofuranones. Organic Letters. 2006;**8**:4637-4640. DOI: 10.1021/ol061924f

[43] Zhang J, Xing C, Tiwari B, Chi YR. Catalytic activation of carbohydrates as formaldehyde equivalents for Stetter reaction with enones. Journal of the American Chemical Society. 2013;**135**: 8113. DOI: 10.1021/ja401511r

[44] Hirano K, Biju AT, Piel I, Glorius F. N-Heterocyclic carbene-catalyzed hydroacylation of unactivated double bonds. Journal of the American Chemical Society. 2009;**131**:14190. DOI: 10.1021/ja906361g

[45] Biju AT, Glorius F. Intermolecular N-heterocyclic carbene catalyzed hydroacylation of arynes. Angewandte Chemie, International Edition. 2010;**49**: 9761-9764. DOI: 10.1002/anie.201005490

[46] Sohn SS, Rosen EL, Bode JW. N-Heterocyclic carbene-catalyzed generation of homoenolates: γ-butyrolactones by direct annulations of enals and aldehydes. Journal of the American Chemical Society. 2004;**126**: 14370-14371. DOI: 10.1021/ja044714b

[47] Burstein C, Glorius F. Organocatalyzed conjugate umpolung of α,β-unsaturated aldehydes for the synthesis of γ-butyrolactones. Angewandte Chemie, International Edition. 2004;**43**:6205-6208. DOI: 10.1002/anie.200461572

[48] Burstein C, Tschan S, Xie X, Glorius F. N-Heterocyclic carbene-catalyzed conjugate umpolung for the synthesis of γ-butyrolactones. Synthesis. 2006:2418-2439. DOI: 10.1055/s-2006-942447

[49] Nair V, Vellalath S, Poonoth M, Mohan R, Suresh E. N-Heterocyclic carbene catalyzed reaction of enals and 1,2-dicarbonyl compounds: Stereoselective synthesis of spiro γ-butyrolactones. Organic Letters. 2006;**8**: 507-509. DOI: 10.1021/ja0625677

[50] Nair V, Vellalath S, Poonoth M, Suresh E, Viji S. N-Heterocyclic carbene catalyzed reaction of enals and diaryl-1,2diones via homoenolate: Synthesis of 4,5,5-trisubstituted γ-butyrolactones. Synthesis 2007, **No. 20**, 3195–3200. DOI: 10.1055/s-2007-990781

[51] Lv H, Jia W-Q, Sun L-H, Ye S. N-Heterocyclic carbene catalyzed [4+3] annulation of enals and o-quinone methides: Highly enantioselective synthesis of benzo-ε-lactones. Angewandte Chemie, International Edition. 2013;**52**:8607-8610. DOI: 10.1002/ange.201303903

[52] Nair V, Poonoth M, Vellalath S, Suresh E, Thirumalai R. An N-heterocyclic carbene-catalyzed [8+3] annulation of tropone and enals via homoenolate. The Journal of Organic Chemistry. 2006;**71**:8964-8965. DOI: 10.1021/jo0615706

[53] Rommel M, Fukuzumi T, Bode JW. Cyclic ketimines as superior electrophiles for NHC-catalyzed homoenolate additions with broad scope and low catalyst loadings. Journal of the American Chemical Society. 2008;**130**: 17266-17267. DOI: 10.1021/ja807937m

[54] He M, Bode JW. Enantioselective, NHC-catalyzed bicyclo-β-lactam formation via direct annulations of enals and unsaturated N-sulfonyl ketimines. Journal of the American Chemical Society. 2008;**130**:418-419. DOI: 10.1021/ja0778592

[55] Chan A, Scheidt KA. Direct amination of homoenolates catalyzed by N-heterocyclic carbenes. Journal of the American Chemical Society. 2008;**130**: 2740-2741. DOI: 10.1021/ja711130p

[56] Seayad J, Patra PK, Zhang Y, Ying JY. Organocatalytic synthesis of N-phenylisoxazolidin-5-ones and a one-pot synthesis of β-amino acid esters. Organic Letters. 2008;**10**:953-956. DOI: 10.1021/ol800003n

[57] Ikota H, Ishida T, Tsukano C, Takemoto Y. Synthesis of 3,3-disubstituted indoline-2-thiones catalyzed by an N-heterocyclic carbene. Chemical Communications. 2014;**50**:8871-8874. DOI: 10.1039/C4CC04047C

[58] Nair V, Vellalath S, Poonoth M, Suresh E. N-Heterocyclic carbene-catalyzed reaction of chalcones and enals via homoenolate: An efficient synthesis of 1,3,4-trisubstituted cyclopentenes. Journal of the American Chemical Society. 2006;**128**:8736-8737. DOI: 10.1021/ja0625677

[59] Nair V, Babu BP, Vellalath S, Varghese V, Raveendran AE, Suresh E. Nucleophilic heterocyclic carbene catalyzed annulation of enals to chalcones in methanol: A stereoselective synthesis of highly functionalized cyclopentanes. Organic Letters. 2009;**11**: 2507-2510. DOI: 10.1021/ol900571x

[60] Ryan SJ, Candish L, Lupton DW. Acyl anion free N-heterocyclic carbene organocatalysis. Chemical Society

Reviews. 2013;**42**:4906-4917. DOI: 10.1039/C3CS35522E

[61] De Sarkar S, Studer A. NHC-catalyzed Michael addition to α,β-unsaturated aldehydes by redox activation. Angewandte Chemie, International Edition. 2010;**49**: 9266-9269. DOI: 10.1002/anie.201004593

[62] Rong Z-Q, Jia M-Q, You S-L. Enantioselective N-heterocyclic carbene-catalyzed Michael addition to α,β-unsaturated aldehydes by redox oxidation. Organic Letters. 2011;**13**: 4080-4083. DOI: 10.1021/ol201595f10

[63] Wanner B, Mahatthananchai J, Bode JW. Enantioselective synthesis of dihydropyridinones via NHC-catalyzed aza-Claisen reaction. Organic Letters. 2011;**13**:5378-5381. DOI: 10.1021/ol202272t

[64] Kravina AG, Mahatthananchai J, Bode JW. Enantioselective, NHC-catalyzed annulations of trisubstituted enals and cyclic N-sulfonylimines via α,β-unsaturated acyl azoliums. Angewandte Chemie, International Edition. 2012;**51**:9433-9436. DOI: 10.1002/anie.201204145

[65] De Sarkar S, Grimme S, Studer A. NHC catalyzed oxidations of aldehydes to esters: Chemoselective acylation of alcohols in presence of amines. Journal of the American Chemical Society. 2010;**132**:1190-1191. DOI: 10.1021/ja910540j

[66] Maji B, Vedachalan S, Ge X, Cai S, Liu X-W. N-Heterocyclic carbene-mediated oxidative esterification of aldehydes: Ester formation and mechanistic studies. The Journal of Organic Chemistry. 2011;**76**:3016-3023. DOI: 10.1021/jo200275c

[67] Sun F-G, Sun L-H, Ye S. N-Heterocyclic carbene-catalyzed enantioselective annulation of bromoenal and 1,3-dicarbonyl compounds. Advanced Synthesis and Catalysis. 2011;**353**:3134-3138. DOI: 10.1002/adsc.201100622

[68] Yetra SR, Bhunia A, Patra A, Mane MV, Vanka K, Biju AT. Enantioselective N-heterocyclic carbene-catalyzed annulations of 2-bromoenals with 1,3-dicarbonyl compounds and enamines via chiral α,β-unsaturated acylazoliums. Advanced Synthesis and Catalysis. 2013;**355**: 1089-1097. DOI: 10.1002/adsc.201300219

[69] Yetra SR, Kaicharla T, Kunte SS, Gonnade RG, Biju AT. Asymmetric N-heterocyclic carbene (NHC)-catalyzed annulation of modified enals with enolizable aldehydes. Organic Letters. 2013;**15**:5202-5205. DOI: 10.1021/ol4026155

[70] Cheng J, Huang Z, Chi YR. NHC organocatalytic formal LUMO activation of α,β-unsaturated esters for reaction with enamides. Angewandte Chemie, International Edition. 2013;**52**: 8592-8596. DOI: 10.1002/anie.201303247

[71] Ryan SJ, Candish L, Lupton DW. N-Heterocyclic carbene-catalyzed generation of α,β-unsaturated acylimidazoliums: Synthesis of dihydropyranones by their reaction with enolates. Journal of the American Chemical Society. 2009;**131**: 14176-14177. DOI: 10.1021/ja905501z

[72] Candish L, Lupton DW. The total synthesis of (−)-7-deoxyloganin via N-heterocyclic carbene catalyzed rearrangement of α,β-unsaturated enol esters. Organic Letters. 2010;**12**: 4836-4839. DOI: 10.1021/ol101983h

[73] Merski M, Townsend CA. Observation of an acryloyl-thiamin diphosphate adduct in the first step of clavulanic acid biosynthesis. Journal of the American Chemical Society. 2007;

129:15750-15751. DOI: 10.1021/ja076704r

[74] Khaleeli N, Li R, Townsend CA. Origin of the β-lactam carbons in clavulanic acid from an unusual thiamine pyrophosphate-mediated reaction. Journal of the American Chemical Society. 1999;**121**:9223-9224

Chapter 9

N-Heterocyclic Carbenes: A Powerful Catalyst for Polymerization

Eman A. Ali

Abstract

As much of the concern is being placed on metal-free polymerization, carbenes are attracting researcher attention. Besides their impact on organic synthesis, they played an integral role in several types of polymerization. This unique class of organocatalyst revives the preparation of polymeric materials that possess functional groups at each carbon atom on the polymer chain, which was hardly reached by conventional methods. This chapter will concern with the contribution of N-heterocyclic carbenes (NHCs) in the preparation of functional polymers from diversified monomers. Also, will discuss the advantages of N-heterocyclic carbenes in some conventional polymerization such as ring-opening and step-growth polymerizations along with the direct zwitterionic polymerization.

Keywords: ring-opening polymerization, step-growth polymerization, zwitterionic polymerization, metal-free polymerization, heterocyclic carbenes

1. Introduction

In the breathtaking development of polymers, the preparation of biocompatible polymers in the economic and environmental friendliness rout has achieved great interests. The synthesis of polymers is usually considered as macromolecular architecture that provides versatile materials in a different application. This includes sophisticated design by different polymerization or post-polymerization techniques [1]. Besides, polymeric materials provided by free radical polymerization, most of the time, the well-known commercial polymers were developed using metal-based catalysts. In this regard, hazardous residual metals and by-products arise as a precarious issue in the biomedical and electronic applications. The cost of precious rare metals, as well as the purification steps, increases the expenditure in large-scale production. In this context, metal-free polymerization by organocatalysts overcomes this obstacle and offers a variety of new synthetic strategies. N-heterocyclic carbenes (NHCs) are classified as one of the most reactive compounds in organocatalysis. In the late 19th and early 20th centuries, NHCs were described as reactive intermediates because the isolation of carbenes was not achieved [2]. N-heterocyclic carbenes have their roots back. Mizuhara et al. [3] reported, in 1954, a natural nucleophilic carbene existence was a catalytically active species of the coenzyme thiamine (**Figure 1**). Ever since the successful isolation of stable NHCs in the early 1990s [2], their contribution has been enlarged rapidly in synthetic chemistry. They

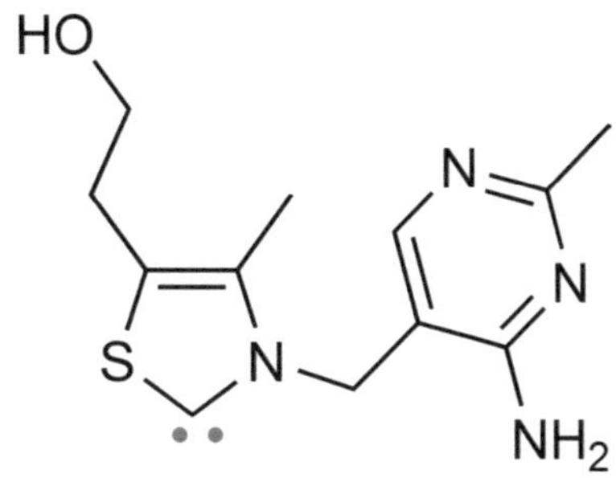

Figure 1.
Coenzyme thiamine.

offer a variety of catalysis and reaction pathways. Besides, their estimated impact on organic synthesis, NHCs are considerable catalysis in the polymer chemist toolbox.

1.1 Polymerization and organocatalysis

In the history of chemistry reactions, catalysis was performed by enzymes and transition metal species. In yet, organocatalysis has emerged to play an integral part in catalysis systems. With regard to other catalytic systems, organocatalysis has been inescapable for many reasons. Beyond their derivation from a variety of organic reagents with plenty of chiral forms, organocatalysis systems are eco-friendly reagents having a low toxicity. Therefore, much of the molecular and macromolecular synthesis relies on it. They were developed to catalysis or initiate polymer synthesis for a variety of sensitive applications like biomedical application, food preserving or packaging, and sophisticated electronic species.

Mainly, polymerization is known to be performed by two categorically mechanisms chain growth and step-growth polymerization. Chain growth polymerization (CGP) is distinguished by the formation of reactive intermediate (anion, cation or radicals) throughout the initiation step. These reactive species transfer the reactive center by reacting with a monomer molecule which is called the propagation stage. The progress of polymer chains is contingent by the continuous reaction of monomer molecules with the formed active center until termination occurs by consuming the active center. In step-growth polymerization (SGP), polymerization starts with the reaction between two molecules that compromise two functional groups. Then another molecule reacts with the formed dimer and so on. Consequently, polymers chain formation depends on the reaction between molecules and/or the formed small chains [4, 5]. Although the difference between these two polymerization mechanisms, they all share using catalytic or initiating systems not only to establish a polymerization process but sometimes to design the macromolecule structure.

Various types of organocatalysts have been employed either in chain growth or step-growth polymerization. It is true that excessive use of organocatalysts was in chain polymerization, in particular, the ring-opening polymerization. However, very recently, many researchers were motivated to use organocatalysts in step-growth polymerization. Given the constantly similar nature of functional groups of ring-opening polymerization, a true example of chain polymerization, with step-growth polymerization, it is nearly to have the same catalytic system for both polymerization mechanisms [6]. Across the field of metal-free polymer preparation catalysis, N-heterocyclic carbenes (NHCs) have affirmed the potential of organocatalysis. This will be presented by revealing NHCs capability to activate certain groups which impact the synthesis of metal-free polymers that are commercially important.

2. N-heterocyclic carbenes (NHCs) overview

Nowadays, developing new polymeric material that possesses biocompatible properties has been strongly emerged. Using organic metal-free catalysts became an inevitable approach in today's environmental mindset. Hence, these catalysts can be easily removed from polymers unlike toxic metals in other types of catalytic systems. Carbenes, in particular N-heterocyclic carbenes (NHCs), are considered as a class of organic metal-free catalysts for different types of the polymerization process. Ever since the first successful isolation of N-heterocyclic carbenes by Arduengo, in the early 1990s, the chemistry richness of these compounds has been revealed in many applications. Their chemical structure can be described as heterocyclic moiety having at least one nitrogen atom and of course carbene carbon [7]. These neutral divalent species of carbon owing only four electrons have participated in σ-bonds and two remained at the central carbon. The presence of nitrogen atoms elevates the stability of carbenes by their ability of π donation to the empty carbon π orbital along with σ withdrawing (**Figure 2**).

This behavior leads to a huge gap of σ-pπ (**Figure 3**) that precedes the strong nucleophilic feature of NHCs. Nevertheless, some NHCs would have amphiphilic character. By substituting the amino with σ-donating alkyl group, an increase of electrophilicity and also nucleophilicity is observed. Also, the incorporation of carbonyl groups into the backbone augmented electrophilicity over the nucleophilicity as they compete with the carbene center for the π donation of the nitrogen atom [8].

Therefore, by studying NHCs ability to donate the electron pair (Lewis basicity) it was found that the triazole-ylidene is less nucleophilic by 10^3 than Imidazole and imidazoline-type (**Figure 3**). Many studies of proton affinity of NHCs, by evaluating the pKa of their conjugated acid, have been employed. They revealed the great impact of the electron-donating substituent on the nitrogen atom as well as the bulkiness of NHCs on their Bronsted basicity. Also, the increase from 5 to 6 membered ring increases the carbenes angles, and leads to an increase in pKa [9–11].

Besides, their distinctive coordination chemistry, N-heterocyclic carbenes have other advantages one of them is they can be easily be modulated bearing in mind the large library of heterocyclic chemistry as shown in **Figure 4**. However, several methods of preparation can be categorized in imidazolium deprotonation, imidazole-thione reduction, and NHCs-adducts thermolysis [12–16].

NHCs have been heavily exploited as ligands for transition metals [17–20]. However, their superiority in metal-free transformations is well recognized in organocatalytic chemistry [21] as well as in macromolecular chemistry [22, 23].

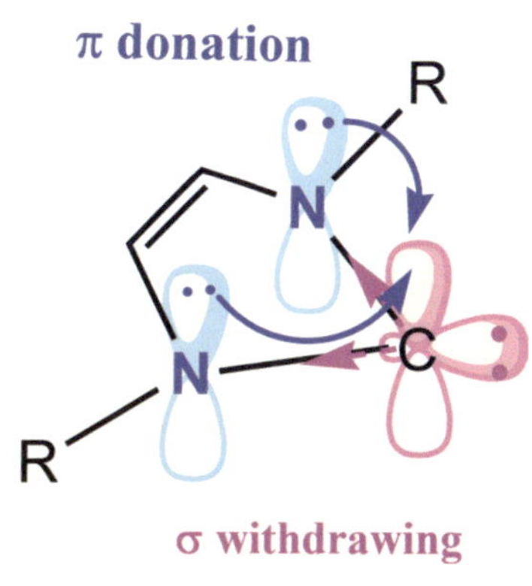

Figure 2.
Ground-state electronic structure of one class of N-heterocyclic carbenes.

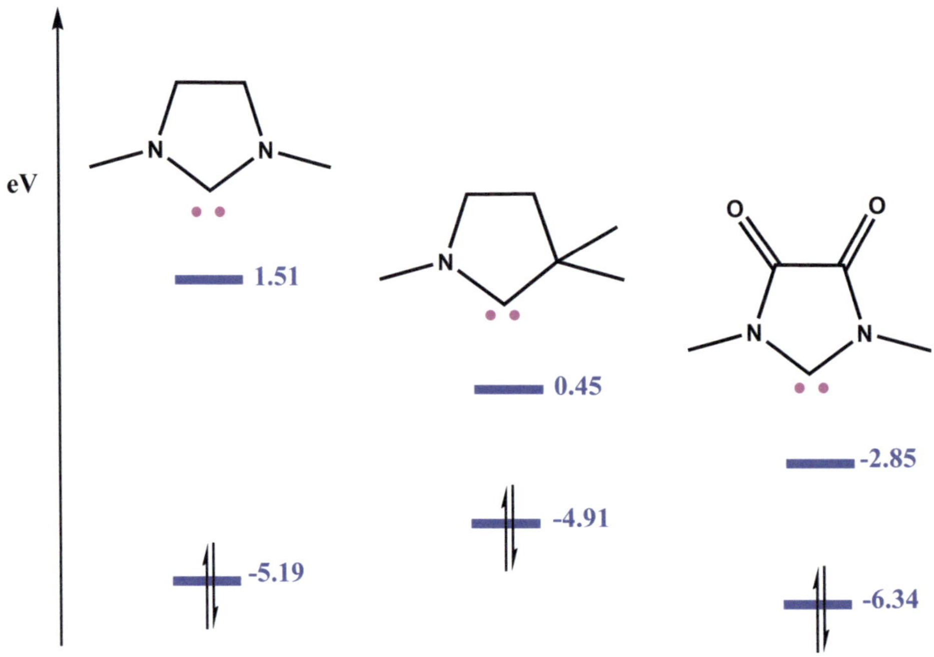

Figure 3.
Energy (eV) of border orbitals of classical NHC.

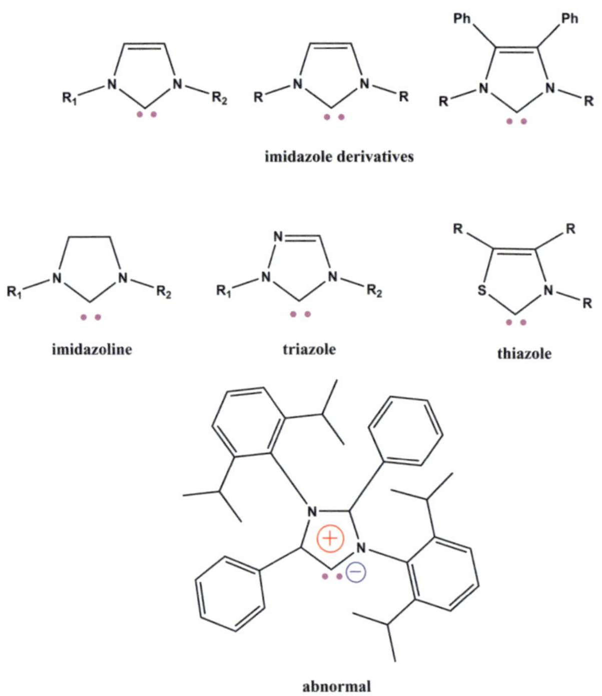

Figure 4.
Examples of N-heterocyclic carbenes polymerization catalysis.

2.1 Ring-opening polymerization (ROP)

Ring-open polymerization has been devoted to developing interesting industrial polymers by synthesis of the analogs of natural as well as biocompatible polymers by different methods. The sharp improvement in ROP is undoubtedly accelerated by organocatalysis. Mainly, organocatalysis of ROP proceeds according to four activation mechanisms; electrophilic monomer activation, nucleophilic monomer activation, base chain-end activation, or bi-functional activation mechanism. Both electrophilic and nucleophilic monomer activation starts by attacking the carbonyl group of the monomer to obtain a macromolecule that bears two ends having opposite charge starts what is called Zwitterionic ROP (ZROP) (**Figure 5**) [24]. However, they differ in their act for activating the carbonyl group. In electrophilic monomer activation, the carbonyl group is activated by protonation or H- bonding attachment that gives room for a chain end nucleophilic attack. While in nucleophilic monomer activation, the zwitterionic intermediate extends a deprotonation process of the alcohol. Then, the formed alkoxide proceeded with the acylation of the carbonyl group. Consequently, the catalyst is free to act again. The third activation mechanism is the chain-end activation where the nucleophilicity of the alcohol is elevated through deprotonation to form either alkoxide or H-bonding [6].

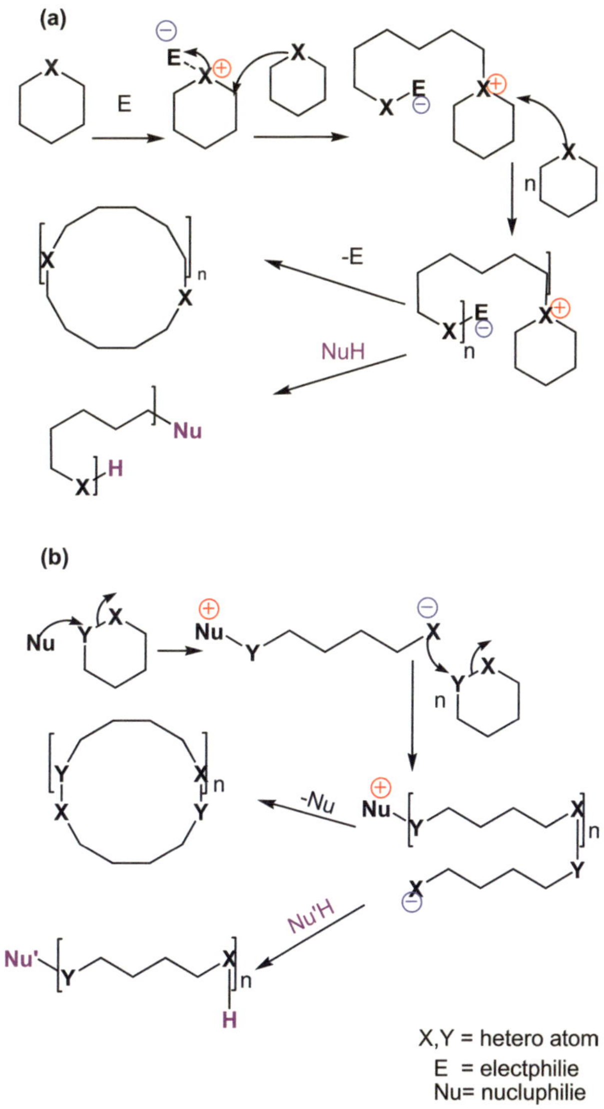

Figure 5.
Mechanism of (a) electrophilic and (b) Nucluphilic zwitterionic ring-opening polymerization.

This chain-end attacks the carbonyl carbon triggering a ring-opening reaction to form an ester allowing the activated alcohol species to reform. The last mechanism for ROP is the bifunctional activation mechanism. It compromises activation of the monomer carbonyl carbon through electrophilic activation along with the activation of the chain end/initiator [25].

Ever since, knowing the benefits of NHCs in transesterification reactions [26–28], they were intensely employed in ring-opening polymerization (ROP). NHCs play a role in producing polymers with low disparities as they are able to provide living polymerization that control the polymer molecular weight. Furthermore, they facilitate the ROP for production of linear and cyclic aliphatic polyesters [29].

2.2 ROP of cyclic ester

Thanks to Nyce et al. in 2002, through their navigation for an efficient nucleophilic catalyst, they discovered the effectiveness of NHCs as organocatalysts for ROP [28]. They also succeeded to polymerize cyclic monomers to deliver Poly (L-lactide) (PLA) (**Figure 6**), poly(ε-caprolactone) (PCL), and poly(b-butyrolactone) (PBL) with dispersity near to unity and definite chain ends which help to control the polymers molecular weight [26]. The polymerization was initiated by alcohols (benzyl alcohol or 4-(pyrene-1-yl)butan-1-ol) which provoke an α-end group address the ester from the initiating alcohol upon ring-opening a hydroxyl functional ω-chain end that propagates the chain. Hedrick's team first suggestion for the transesterification reaction mechanism was activated monomer mechanism. Considering the steric effect and the higher pKa of the alcohol compared to the conjugated acid of NHC in DMSO, deprotonation of less acidic alcohol by NHC is unlikely the beginning step of the catalysis act. Therefore, they assumed a direct attack of the monomer by the nucleophilic NHC to form a zwitterionic intermediate that interacts with the other monomer molecules pursued by the reaction with alcohol. Another initiation mechanism proposed by the theoretical study assumed the occurrence of an active chain-end mechanism. Lia *et al.* suggest the hydrogen bonding between NHC and alcohol, then deprotonation of alcohol which attacks the cyclic monomer [30]. This assumption was based on the lower energy of the H-bonded adduct than the zwitterionic intermediate. Several studies follow to

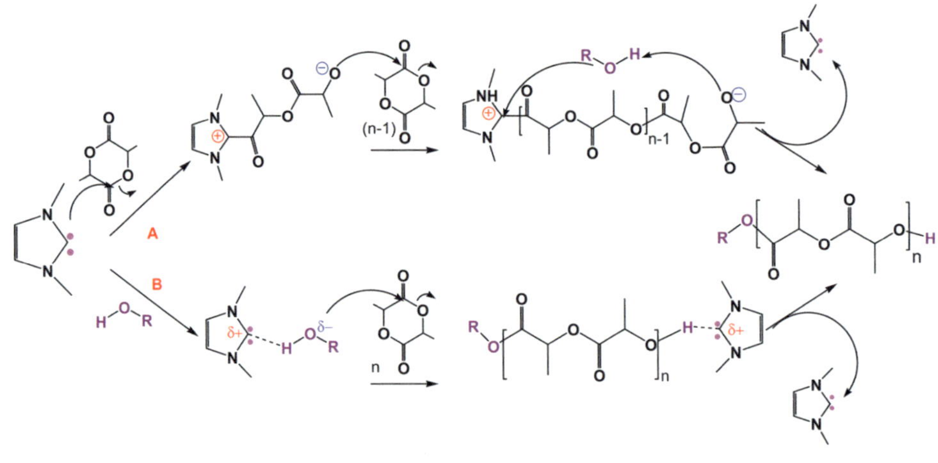

Figure 6.
Ring-opening polymerization of L-lactide through path: (A) monomer activation mechanism and (B) active-chain end mechanism.

find out the predominant mechanism. This dispute most likely has been resolved by Patel *et al.* manifesting the ability of NHC to act as bifunctional catalysts in the presence of alcohol and the two mechanisms are likely participating in [31].

The catalytic behavior NHCs in the absence of alcohol was investigated. At a relatively high LA concentration and ambient temperatures, a very fast polymerization was reported (5 s–900 s) yielding a cyclic polymer. In this case, NHC acts as an initiator that generates zwitterionic intermediate by a direct nucleophilic attack of NHC to the LA monomer. The ring-closure occurred by trapping the NHC within a zwitterionic NHC–CS_2 adduct.

Engaging the spirit of the suggested mechanism of cyclic esters polymerization, remarkable turnovers were observed for the ROP of a variety of other cyclic monomers including cyclosiloxanes, epoxides, and N-carboxyanhydrides. NHCs proved extreme activeness, although the usage of low concentration and temperature.

2.3 ROP of siloxanes

Taking the advantage of NHCs silicophilicity [32], the ROP of cyclic (carbo) siloxanes has been investigated [33]. A rapid polymerization of 2,2,5,5-tetramethyl-1-oxa-2,5-disilacyclopentane (TMOSC) occurred in less than a minute [34]. The polymerization was activated by electron-rich NHC in toluene (**Figure 7**). The product, poly(carbosiloxane), the molar mass of 10,200 g mol^{-1} was controlled with dispersity equal to 1.19. This was observed if the polymerization stopped in high conversion otherwise broadening was detected due to undesired transesterification side reactions.

The authors reported a decrease in the polymerization rate when bulky, and less basic NHC is involved. Also, they revealed through mechanistic studies that the polymerization process is activated by hydrogen bonding instead of nucleophilic ring-opening of TMOSC by the NHC.

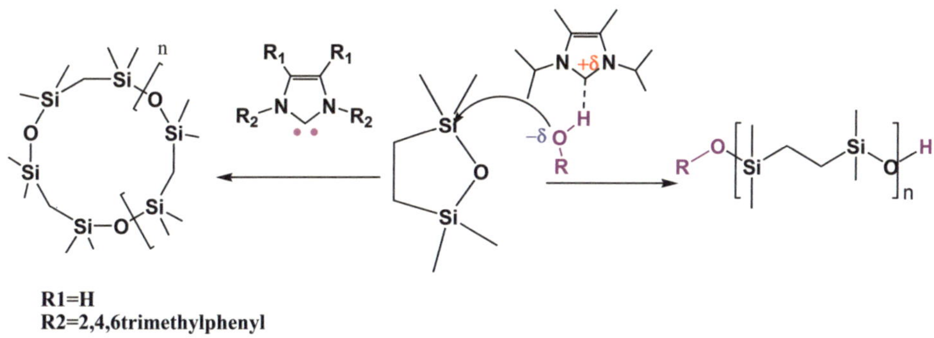

Figure 7.
the proposed mechanism of ROP of TMOSC.

2.4 ROP of epoxides

An attempt to activate the ROP of ethylene oxide (EO) by NHC was recorded by Raynaud et al. [34]. In this work, NHC succeeded to accelerate the ROP of ethylene oxide as a direct initiator and combined with chain regulators of the NuE-type. 1,3-diisopropylimidazol-2-ylidene initiate alone ROP of EO in DMSO at 50°C. linear difunctionalized PEOs were produced, unlike cyclic polymers that formed by ZROP of LA which was previously discussed.

2.5 ROP of carbonates

The ROP of cyclic carbonates was reported in the melt or bulk polymerization [34]. Nederberg et al. investigated the ring-opening polymerization of trimethylene carbonate using varieties of organocatalysts. NHC catalysis yield a quantitative conversion in 30 min and a polymer dispersity of only 1.06, when less electron-rich NHC is employed (**Figure 8**). This study showed that the increased electron-rich nature of NHC leads to an increase in the polymerization rate at the expense of molecular weight control (99% conversion in seconds).

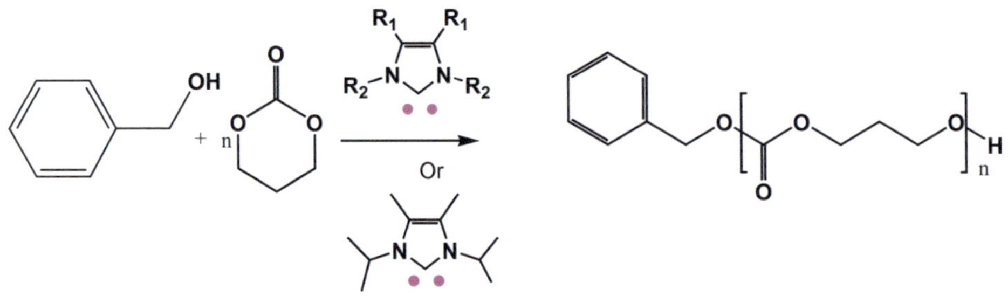

Figure 8.
The ring-opening polymerization of trimethylene carbonate.

2.6 ROP of anhydride

N-carboxyanhydrides have been polymerized using NHC to produce linear poly(α-peptoids) in THF [35]. One of the biggest advantages of this polymerization is the ability to prepare a definite structure with low molecular weight distributions in the range of 1.04–1.12and the molecular weight ranges (3000–40,000 gmol^{-1}). The authors found that small N-substituents of NHC enhance the reaction rate. They also revealed that the control of molecular weight is strongly dependent on the solvent and the NHC structure. The mechanism of the polymerization followed the ROP mechanism under the loss of CO_2. Side reactions are significantly suppressed in low dielectric solvents due to the reduced basicity and nucleophilicity of the negatively charged chain ends of the zwitterions, resulting in quasi-living polymerization behavior.

2.7 Step-growth polymerization

Virtually all high-performance polymers (80%) that are currently utilized are products of chain-growth polymerization along with step-growth polymerization. The top valued polymers, polyether ketones, polysulfones, polyimides are step-growth polymerization products. Normally, step-growth polymerization (SGP) compromises the reaction between two different bi-functional groups that might present in one monomer or two different monomers. Amidation, esterification, nucleophilic aromatic substitution, transesterification, and urethane formation with isocyanates are the conventional reaction in step-growth polymerization. They almost proceed with the high conversion that is suitable for polymerization. However, the hard condition, high pressures and temperatures, and side reaction leading to monomers decomposition and limiting the molecular weight [6]. Therefore, almost all step-growth polymerizations require a catalyst to increase the rate of reactions and consequently reduce the potential side reactions.

(NHCs) have been used in step-growth polymerization to achieve high molecular weight polymers. Mostly, they were in-situ developed through deprotonation of imidazolium salts with a base.

Bearing in mind their potential in transesterification reaction, NHCs catalysis was implemented in step-growth polymerization of 6-hydroxyhexanoate, bis(2-hydroxyethyl) terephthalate as well as the polycondensation of dimethylcarbonate (DMC) and a number of diols.

Hedrick et al. polymerized bis(2-hydroxyethyl) terephthalate using only NHCs as a catalytic agent in THF. The polymerization process accomplished almost full conversion within one hour at 250°C. They also, succeeded to prepare aliphatic polyesters by polytransesterification reactions of ethyl 6-hydroxyhexanoate and ethyl glycolate [28]. Poly-(6-hydroxyhexanoate) with dispersity of 1.57 and Mn of 21,000 gmol^{-1}was obtained by carrying out the SGP at 60°C for 24 h. The polymer in 95% yield was obtained by removing EtOH at low pressure. By this procedure, polyesters (with Mn ranging from 8000 to 20,000 gmol^{-1}) were similar to poly(ε-caprolactone) (PCL) and poly-(glycolide) synthesized by ring-opening polymerization (ROP).

NHCs activate the monomers by attacking their carbonyl carbon. This feature was also implemented to prepare a variety of industrial polymers. Plasseraud et al. reported their success to prepare metal-free aliphatic polycarbonates [36]. Dimethylcarbonate and diols in molar mass equal 3:1, respectively, were reacted in bulk at 150°C under reduced pressure. The reactions were conducted at 100°C for 15 min in the first stage to liberate the active NHC by decarboxylation of the NHC–CO_2 adduct that was used as precatalyst. Thereafter, the temperature was elevated to 150°C for one hour under reduced pressure to remove methanol which forceful the polymer formation. Random copolymer with moderately controlled molecular weight distributions and molecular weight (19,000 gmol^{-1}) and homopolymers were produced. Employing a molar equivalent 1: 2 of DMC and aliphatic diols, respectively, hydroxy-terminated polycarbonates could also be achieved.

Umpolung reactions have their influence on polymer chemistry. The benzoin condensation reaction motivated Pinaud, et al. to synthesis polybenzoin [37]. In this case, the carbonyl group in bis-aldehyde is activated by NHC in THF or DMSO at 40°C to form alkoxide that triggers the formation of "Breslow intermediate". This intermediate attack the electrophilic carbon of another aldehyde molecule (**Figure 9**). Thereafter, C-C bond formation leads to the step-growth polymerization of bis-aldehyde and cyclic polymers by-products.

In another pathway, NHCs have been used for activation of the alcohol for developing interesting polyurethane (PU) from isocyanates and polyols reaction. A study performed by Bantu et al. showed that the order of addition is a key for successful formation of PU [38, 39]. Hence, first, the alcohol was deprotonated

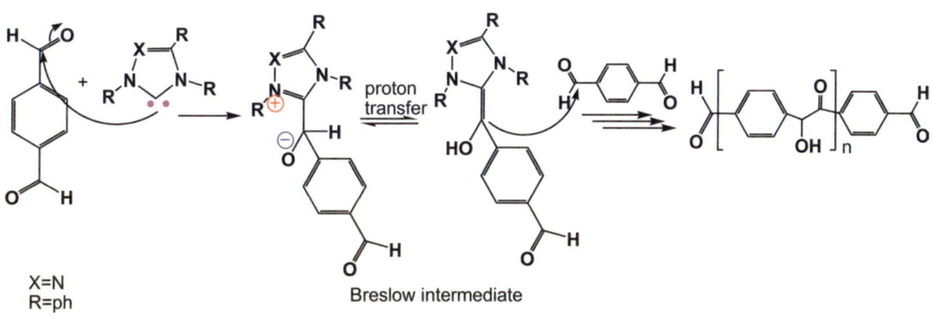

Figure 9.
The proposed mechanism of the step-growth polymerization of bis-aldehyde.

by the NHC before the addition of the di-isocyanate monomer. In this investigation, the synthesis of cross-linked polyurethanes was conducted in CH_2Cl_2 at 60–70°C affording in-situ generation of NHC catalyst from NHC–CO_2 adducts. The resulting alkoxides from the reaction of NHC catalyst and ethylene glycol or polyol in a 1/1 ratio at 70°C were detected quantitatively by 1H NMR analysis. The C2H imidazolium proton and pyridinium proton were detected confirming the proposed mechanism of alcohol activation. Not only the order of addition of reactants is vital but also the nature of the diisocyanate monomer. Coutelier et al., found that when linear aliphatic diisocyanates are employed, soluble, linear PUs (2000–5000 $gmol^{-1}$) might be derived [40] otherwise crosslinked PU is formed. The SGP polymerizations were carried out in THF using 1 mol% catalyst relative to monomer between 30 and 50°C. The 1/1 ratio was employed for a selected diol and two aliphatic diisocyanates (isophorone diisocyanate and 1,6-diisocyanatohexane). Despite the potency of NHCs as catalysts for the cyclo di or trimerization reaction of phenyl monoisocyanate (70% cyclodimer and 30% cyclotrimer) [41], traces of such uretdione or isocyanurate were detected with alkyl isocyanates. This provides another confirmation of the alcohol activation through H-bonding before nucleophilic addition onto the isocyanate species.

This activation mechanism was utilized by Marrot et al. for the polycondensation of disilanols [42]. In a closed schlenk tube, 揬α,ω-Dihydroxy oligodimethylsiloxanes was mixed with a catalytic amount of isolated NHCs at 80°C for 16 h to yield almost 90%. Interestingly, the water released from the dehydration of the silence did not depress the catalytic activity of NHC. The hydrophobic nature of the developed polydimethylsiloxane seems to prevent direct contact with NHC. Nevertheless, removing the produced water leads to increasing molecular weights of the resulting silicone polymers. This observation suggests another role for NHCs as a catalyst for depolymerization reactions in the presence of H_2O. Therefore, the catalytic amount of NHC and water withdrawal have an effect on regulating the produced polymer molecular weight.

3. Conclusions

Throughout the past two decades N-heterocyclic carbenes (NHCs), have well stood as a true organocatalyst for the production of many industrial polymers. Owing to their rich structural modularity, NHCs can afford highly selective polymerization reaction pathways. A deep awareness of NHC's catalytic activity potential was gained through understanding their activation reaction mechanism that opens pathways for the production of commercial polymers. They have been extensively involved as transesterification agents in the ROP. Also, they showed a tremendous impact on step-growth reactions for the production of high molecular weight polymers (polycarbonates, polyesters, polybenzoins). Besides, their role of accelerating polymerization and their temperature range extends, they have the ability to introduce functionality to polymers. Due to their sensitivity to air and moisture, NHCs were in-situ generated using affordable and air-stable precursors, imidazolium chloride salt as starting source. As the catalyst design field progresses, opportunities for NHC polymerization catalysis can move beyond its current niche to compete in a field currently dominated by heterogeneous metal catalysis.

Author details

Eman A. Ali
Polymers and Pigments Department, National Research Centre, Giza, Egypt

*Address all correspondence to: emabal2003@yahoo.com

References

[1] Hu S, Zhao J, Zhang G, et al. Macromolecular architectures through organocatalysis. Progress in Polymer Science. 2017;**74**:34-77

[2] Enders D, Niemeier O, Henseler A. Organocatalysis by N-heterocyclic carbenes. Chemical Reviews. 2007;**107**:5606-5655

[3] Mizuhara S, Tamura R, Arata H. On the mechanism of thiamine action II. Proc Jpn Acad. 1951;**27**:302-308

[4] Odian G. Principles of polymerization. New Jersey, Printed in the United States of America: John Wiley & Sons; 2004

[5] Koltzenburg S, Maskos M, Nuyken O. Industrially relevant polymerization processes. In: Polymer Chemistry. Berlin, Heidelberg: Springer; 2017. DOI: 10.1007/978-3-662-49279-6_16

[6] Bossion A, Heifferon KV, Meabe L, et al. Opportunities for organocatalysis in polymer synthesis via step-growth methods. Progress in Polymer Science. 2019;**90**:164-210

[7] Hopkinson MN, Richter C, Schedler M, et al. An overview of N-heterocyclic carbenes. Nature. 2014;**510**:485-496

[8] Martin D, Lassauque N, Donnadieu B, et al. A cyclic diaminocarbene with a pyramidalized nitrogen atom: A stable N-heterocyclic carbene with enhanced electrophilicity. Angew Chemie-Int Ed. 2012;**51**:6172-6175

[9] César V, Lugan N, Lavigne G. A Stable Anionic N-heterocyclic carbene and its zwitterionic complexes. Journal of the American Chemical Society. 2008;**130**:11286-11287

[10] Magill AM, Cavell KJ, Yates BF. Basicity of nucleophilic carbenes in aqueous and nonaqueous solvents theoretical predictions. Journal of the American Chemical Society. 2004;**126**: 8717-8724

[11] Higgins EM, Sherwood JA, Lindsay AG, et al. p K as of the conjugate acids of N-heterocyclic carbenes in water. Chemical Communications. 2011;**47**:1559-1561

[12] Nyce GW, Csihony S, Waymouth RM, et al. A general and versatile approach to thermally generated N-heterocyclic carbenes. Chemistry - A European Journal. 2004;**10**:4073-4079

[13] Benhamou L, Chardon E, Lavigne G, et al. Synthetic routes to N-heterocyclic carbene precursors. Chemical Reviews. 2011;**111**:2705-2733

[14] Fliedel C, Braunstein P. Recent advances in S-functionalized N-heterocyclic carbene ligands: From the synthesis of azolium salts and metal complexes to applications. Journal of Organometallic Chemistry. 2014;**751**:286-300

[15] Jahnke MC, Hahn FE. Introduction to N-heterocyclic carbenes: Synthesis and stereoelectronic parameters. In: N-Heterocyclic Carbenes: From Laboratory Curiosities to Efficient Synthetic Tools. Edition (2). London, UK: The Royal Society of Chemistry; 2016. pp. 1-45

[16] Vellé A, Cebollada A, Macías R, et al. From Imidazole toward Imidazolium Salts and N-Heterocyclic Carbene Ligands: Electronic and Geometrical Redistribution. ACS Omega. 2017;**2**:1392-1399

[17] Suzuki Y, Ono RJ, Ueda M, et al. N-heterocyclic carbene enabled synthesis of conjugated polymers. European Polymer Journal. 2013;**49**:4276-4280

[18] Herrmann WA, Kocher C. N-Heterocyclic carbenes. Angew Chemie Int Ed English. 1997;**36**: 2162-2187

[19] Koy M, Bellotti P, Das M, et al. N-Heterocyclic carbenes as tunable ligands for catalytic metal surfaces. Nature Catalysis. 2021;**4**:352-363

[20] Jain I, Malik P. N-heterocyclic carbene complexes in ring opening polymerization. European Polymer Journal. 2021;**150**:110412

[21] Nair V, Bindu S, Sreekumar V. N-heterocyclic carbenes: Reagents, not just ligands! Angew Chemie-Int Ed. 2004;**43**:5130-5135

[22] Fèvre M, Pinaud J, Gnanou Y, et al. N-Heterocyclic carbenes (NHCs) as organocatalysts and structural components in metal-free polymer synthesis. Chemical Society Reviews. 2013;**42**:2142-2172

[23] Naumann S, Dove AP. N-Heterocyclic carbenes for metal-free polymerization catalysis: An update. Polymer International. 2016;**65**:16-27

[24] Coulembier O. Chapter 1: nucleophilic catalysts and organocatalyzed zwitterionic ring-opening polymerization of heterocyclic monomers. In: Organic Catalysis for Polymerisation. London, UK: The Royal Society of Chemistry; 2019. pp. 1-36. DOI: 10.1039/9781788015738-00001

[25] Xia Y, Boydston AJ, Yao Y, et al. Ring-expansion metathesis polymerization: Catalyst-dependent polymerization profiles. Journal of the American Chemical Society. 2009;**131**:2670-2677

[26] Connor EF, Nyce GW, Myers M, et al. First example of N-heterocyclic carbenes as catalysts for living polymerization: Organocatalytic ring-opening polymerization of cyclic esters. Journal of the American Chemical Society. 2002;**124**:914-915

[27] Grasa GA, Kissling RM, Nolan SP. N-heterocyclic carbenes as versatile nucleophilic catalysts for transesterification/acylation reactions. Organic Letters. 2002;**4**:3583-3586

[28] Nyce GW, Lamboy JA, Connor EF, et al. Expanding the catalytic activity of nucleophilic N-heterocyclic carbenes for transesterification reactions. Organic Letters. 2002;**4**:3587-3590

[29] Kiesewetter MK, Shin EJ, Hedrick JL, et al. Organocatalysis: Opportunities and challenges for polymer synthesis. Macromolecules. 2010;**43**:2093-2107

[30] Lai C-L, Lee HM, Hu C-H. Theoretical study on the mechanism of N-heterocyclic carbene catalyzed transesterification reactions. Tetrahedron Letters. 2005;**46**:6265-6270

[31] Patel D, Liddle ST, Mungur SA, et al. Bifunctional yttrium (III) and titanium (IV) NHC catalysts for lactide polymerisation. Chemical Communications. 2006;**10**:1124-1126

[32] Bonnette F, Kato T, Destarac M, et al. Encapsulated N-heterocyclic carbenes in silicones without reactivity modification. Angew Chemie Int Ed. 2007;**46**:8632-8635

[33] Rodriguez M, Marrot S, Kato T, et al. Catalytic activity of N-heterocyclic carbenes in ring opening polymerization of cyclic siloxanes. Journal of Organometallic Chemistry. 2007;**692**: 705-708

[34] Lohmeijer BGG, Dubois G, Leibfarth F, et al. Organocatalytic living ring-opening polymerization of cyclic carbosiloxanes. Organic Letters. 2006;**8**:4683-4686

[35] Guo L, Lahasky SH, Ghale K, et al. N-Heterocyclic carbene-mediated

zwitterionic polymerization of N-substituted N-carboxyanhydrides toward Poly(α-peptoid)s: kinetic, mechanism, and architectural control. Journal of the American Chemical Society. 2012;**134**:9163-9171

[36] Naik PU, Refes K, Sadaka F, et al. Organo-catalyzed synthesis of aliphatic polycarbonates in solvent-free conditions. Polymer Chemistry. 2012;**3**:1475-1480

[37] Pinaud J, Vijayakrishna K, Taton D, et al. Step-growth polymerization of terephthaldehyde catalyzed by N-heterocyclic carbenes. Macromolecules. 2009;**42**:4932-4936

[38] Bantu B, Pawar G, Decker U, et al. CO2 and SnII adducts of N-heterocyclic carbenes as delayed-action catalysts for polyurethane synthesis. Chemistry - A European Journal. 2009;**15**:3103

[39] Bantu B, Manohar Pawar G, Wurst K, et al. CO2, magnesium, aluminum, and zinc adducts of N-heterocyclic carbenes as (latent) catalysts for polyurethane synthesis. European Journal of Inorganic Chemistry. 2009;**2009**:1970-1976

[40] Coutelier O, El Ezzi M, Destarac M, et al. N-Heterocyclic carbene-catalysed synthesis of polyurethanes. Polymer Chemistry. 2012;**3**:605-608

[41] Duong HA, Cross MJ, Louie J. N-Heterocyclic carbenes as highly efficient catalysts for the cyclotrimerization of isocyanates. Organic Letters. 2004;**6**:4679-4681

[42] Marrot S, Bonnette F, Kato T, et al. N-Heterocyclic carbene-catalyzed dehydration of α,ω-disilanol oligomers. Journal of Organometallic Chemistry. 2008;**693**:1729-1732